The Valve Primer

The Valve Primer

Brent T. Stojkov

Industrial Press Inc.

Library of Congress Cataloging-in-Publication Data

Stojkov, Brent T.
    The valve primer / Brent T. Stojkov.
    240 p.
    Includes index.
    ISBN 0-8311-3077-6
    1. Valves. I. Title.
IN PROCESS
621.8'4 — dc21   96-37133
                                                                   CIP

Copyright, 1997 © by Industrial Press Inc., New York, N.Y. 10016. All rights reserved. This book, or parts thereof, may not be reproduced in any form without permission of the publishers.

Printed in the United States of America

4   5   6   7   8   9

Contents

PREFACE		v
ACKNOWLEDGMENTS		vii
REGISTERED TRADEMARKS		viii
LIST OF FIGURES		ix
1	INTRODUCTION	1
2	GATE VALVES	5
3	GLOBE VALVES	25
4	CHECK VALVES	37
5	BUTTERFLY VALVES	55
6	BALL VALVES	65
7	PLUG VALVES	79
8	DIAPHRAGM VALVES	93
9	VALVE MATERIALS	101
10	SIZES, CLASSES, AND RATINGS	121
11	FLUID FLOW THROUGH VALVES	131
12	OPERATORS AND ACTUATORS	141
13	CONTROL VALVES AND PRESSURE RELIEF VALVES	163
14	SELECTION	173
15	MAINTENANCE AND REPAIR	181
16	MISCELLANEOUS TOPICS	189
APPENDIX 1	STANDARDS	201
APPENDIX 2	GLOSSARY	205

Preface

Several years ago I was asked if I knew of a book about valve basics that could be used in training new operators at the oil refinery where I was working at the time. When I replied that I did not, it was suggested that I write one. This book is the result of that suggestion. It is in large part a compilation of information scattered throughout different valve manufacturers' catalogs and design guides, but it also includes knowledge that I have accumulated in over 25 years of experience in valve sales, selection, specification, procurement, inspection, troubleshooting, and repair.

The Valve Primer is written for those individuals in the power generation, oil, chemical, paper, and other processing industries whose jobs as engineers, operators, and maintenance technicians bring them into daily contact with the hundreds to thousands of valves that are found in the power stations, refineries, plants, and mills in which they work. It is a primer in the true sense of the word: a small introductory work. Its objective is not to produce valve experts, but individuals who have a basic knowledge of valve types and designs, the materials valves are made of, where they should and should not be used, when and how they are actuated, and other related topics. Just as important is the reader's introduction to valve terminology. Without a knowledge of valve terms, useful discussion with supervisors, coworkers, valve salesmen, and valve servicemen is greatly hindered.

Many manufacturers of valves and valve actuators have been helpful in allowing catalog photographs of their products to be used as illustrations in *The Valve Primer*. I express my thanks for their generosity. I have also had favorable experiences with products made by many manufacturers that are not represented here; therefore, omission of manufacturers' names does not mean that their products are unsatisfactory.

Brent T. Stojkov
Loveland, Ohio

Acknowledgments

We are indebted to the following organizations for their contributions to *The Valve Primer*.

The American Society of
  Mechanical Engineers
New York, NY

Auma Actuators, Inc.
Pittsburgh, PA

Bettis Actuators & Controls
Waller, TX

DeZurik, Inc.
Sartell, MN

J.M. Huber Corporation
Tomball, TX

ITT Engineered Valves
Lancaster, PA

Instrument Society of
  America
Research Triangle Park, NC

Kunkle Valve Division
Black Mountain, NC

Manufacturers
  Standardization Society
Vienna, VA

Neles-Jamesbury, Inc.
Worcester, MA

Stockham Valves &
  Fittings
Birmingham, AL

TECHNAFLOW Inc.
Vancouver, WA

Velan, Inc.
Montreal, Canada

Henry Vogt Machine Co.
Louisville, KY

The Walworth Co.
Houston, TX

Xomox Corporation.
Cincinnati, OH

Registered Trademarks

The following registered trademarks are used to identify specific materials in this book.

"Fluorel" is a registered trademark of:
  3M
  3M Center, Building 223-6S-04
  St. Paul, MN 55144

"Nordel," "Teflon," "Hypalon," "Tefzel" and "Viton" are registered trademarks of:
  DuPont Co.
  1007 Market Street
  Wilmington, DE 19898

"Hastelloy" and "Hastelloy C" are registered trademarks of:
  Haynes International, Inc.
  1020 W. Park Avenue
  Kokomo, IN 46904

"Inconel" and "Monel" are registered trademarks of:
  INCO Alloys International, Inc.
  3200 Riverside Drive
  Huntington, WV 25705

"Stellite" is a registered trademark of:
  Stoody Co.
  P.O. Box 9997
  Bowling Green, KY 42102

List of Figures

Chapter 2

FIGURE 2-1	Gate Valve *courtesy of The Walworth Company*
FIGURE 2-2	Screwed Body–Bonnet Joint, ISRS *'extracted from MSS SP-80, 1987, with permission of the publisher, the Manufacturers Standardization Society.'*
FIGURE 2-3	Union Body–bonnet Joint *courtesy of The Walworth Company*
FIGURE 2-4	Pressure-Seal Body–bonnet Joint *courtesy of Velan, Inc.*
FIGURE 2-5	Welded Body–bonnet Joint *courtesy of Velan, Inc.*
FIGURE 2-6	Non-rising Stem *'extracted from MSS SP-80, 1987, with permission of the publisher, the Manufacturers Standardization Society.'*
FIGURE 2-7	Section Through Yoke Nut and Stuffing Box
FIGURE 2-8	Wedge Gate Designs
FIGURE 2-9	Knife Gate Valve *courtesy of TECHNAFLOW, Inc.*
FIGURE 2-10	Knife Gate Valve Seat Designs

Chapter 3

FIGURE 3-1	Globe Valve *courtesy of The Walworth Company*
FIGURE 3-2	Globe Valve Disc Designs
FIGURE 3-3	Angle Valve *courtesy of The Walworth Company*
FIGURE 3-4	Y-pattern Globe Valve *courtesy of Velan, Inc.*
FIGURE 3-5	Needle Valve

Chapter 4

FIGURE 4-1	Swing Check Valve *courtesy of The Walworth Company*
FIGURE 4-2	Swing Check Valve with Lever and Weight
FIGURE 4-3	Tilting-Disc Check Valve *courtesy of Velan, Inc.*
FIGURE 4-4	Wafer Check Valve *courtesy of Stockham*

List of Figures

FIGURE 4-5	Lift Check Valve *courtesy of Velan, Inc.*
FIGURE 4-6	Lift Check Valve Disc Guides *courtesy of The Walworth Company*
FIGURE 4-7	Ball Check Valve
FIGURE 4-8	Stop-Check Valve

Chapter 5

FIGURE 5-1	Lined Butterfly Valve *courtesy of DeZurik*
FIGURE 5-2	High-Performance Butterfly Valve *courtesy of Xomox Corporation*
FIGURE 5-3	High-Performance Valve Disc Offsets

Chapter 6

FIGURE 6-1	Split-Body Ball Valve *courtesy of Velan, Inc.*
FIGURE 6-2	End-Entry Ball Valve *courtesy of Velan, Inc.*
FIGURE 6-3	Top-Entry Ball Valve *courtesy of Velan, Inc.*
FIGURE 6-4	3-piece Ball Valve *courtesy of Neles-Jamesbury, Inc.*
FIGURE 6-5	Trunnion Ball Valve *courtesy of Velan, Inc.*
FIGURE 6-6	3-Way Ball Valve *courtesy of Neles-Jamesbury, Inc.*

Chapter 7

FIGURE 7-1	Plug Cock
FIGURE 7-2	Lubricated Plug Valve *courtesy of J.M. Huber Corporation*
FIGURE 7-3	Sleeved Plug Valve *courtesy of Xomox Corporation*
FIGURE 7-4	Multi-port Plug Valve Designs *courtesy J.M. Huber Corporation*
FIGURE 7-5	Multi-port Plug Valve Applications

Chapter 8

FIGURE 8-1	Straight-through Diaphragm Valve *courtesy of ITT Engineered Valves*
FIGURE 8-2	Weir Diaphragm Valve *courtesy of ITT Engineered Valves*

List of Figures

Chapter 10

FIGURE 10-1 Pressure-temperature Ratings of Carbon Steel *courtesy of ASME*
FIGURE 10-2 Pressure-temperature Rating of a Class 150 Ball Valve
FIGURE 10-3 Pressure-temperature Ratings of a Diaphragm Valve

Chapter 11

FIGURE 11-1 Valve Flow Characteristic Curves

Chapter 12

FIGURE 12-1 Spur Gear Operator *courtesy of Auma Actuators, Inc.*
FIGURE 12-2 Bevel Gear Operator *courtesy of Auma Actuators, Inc.*
FIGURE 12-3 Worm Gear Operator *courtesy of Auma Actuators, Inc.*
FIGURE 12-4 Piston Actuator *courtesy of Bettis Corporation*
FIGURE 12-5 Diaphragm Actuator *courtesy of ITT Engineered Valves*
FIGURE 12-6 Rotary Actuator *courtesy of Xomox Corporation*
FIGURE 12-7 Rack-and-Pinion Actuator *courtesy of Bettis Corporation*
FIGURE 12-8 Scotch-yoke Actuator *courtesy of Bettis Corporation*
FIGURE 12-9 Electric Actuator *courtesy of Auma Actuators, Inc.*

Chapter 13

FIGURE 13-1 Block Diagram of a Regulator
FIGURE 13-2 Safety Valve
FIGURE 13-3 Relief Valve *courtesy of Kunkle Valve Division*

Chapter 16

FIGURE 16-1 Valve Bypass
FIGURE 16-2 Extended Body Valve *courtesy of Henry Vogt Machine Co.*
FIGURE 16-3 Bellows Seal Valve

1

Introduction

A valve is a mechanical device whose function is to control the flow of fluids in piping systems. The fluids controlled in industrial settings can be the common liquids, gases, and vapors; but they also can be liquids carrying suspended solid particles (called *slurries*) and gases carrying suspended solid particles. In some instances, even a dry powder can be considered and handled as a fluid. The control applied to these fluids can take one or more of the following forms:

1. Starting and stopping flow
2. Regulating flow volume (frequently called *throttling*)
3. Preventing reverse flow (called *anti-backflow*)
4. Changing flow direction
5. Limiting fluid pressure

A valve performs its control function by placing an obstruction (hereafter called the *flow control element*) in the fluid path through the valve. The nature of the flow control element determines the valve type and the form of control for which the valve is suited. There are many types of valves; however, those used in industrial and power piping applications are almost always one of the following valve types: gate, globe, check, butterfly, ball, plug, or diaphragm. Variations in the designs of each of these valve types have been developed to satisfy many different applications. The flow control element, the form of control, the different designs, and the applications for each of these types are discussed in detail in the subsequent chapters. First, some additional comments about valves in general are appropriate.

In the process of performing its control function, a valve must satisfy two conditions. First, the fluid cannot be allowed to leak into the environment. Second, there can be no internal leakage. That is, when the valve is closed there can be no flow, either along the normal fluid path or between valve parts, where flow is never intended. These two conditions are not always met absolutely. Valve testing standards, such as the American Petroleum Institute Standard API 598, permit a very small amount of leakage at the seating surfaces for some types and sizes of valves. In addition, external leakage past stem packing and shaft seals is not uncommon in valves that have been in service for some time.

All valve types are manufactured with ends that mate with the common piping connection methods: threaded (also called *screwed*), flanged, butt-weld, socket-weld, solder, and grooved. The appropriate connection method to be used on the valves in a specific pipeline is decided by the piping designer. The designer's decision is based on factors such as line size, fluid pressure, materials of construction, and ease of assembly. However, some standardization exists. Small valves are usually manufactured with threaded or socket-weld ends; whereas large valves are usually manufactured with flanged or butt-weld prepared ends. The threaded and flanged connections permit easy assembly and disassembly and must be used where the valve material cannot be welded. However, fluid leakage at these connections is possible. Socket-weld and butt-weld joints are leak-tight and are preferable for high-pressure pipelines; however, they are more difficult to make and are much more permanent than are threaded and flanged joints. Note that the commonly accepted convention is that piping 2 inch NPS (nominal pipe size) and smaller is referred to as *small* or *small bore,* and piping 2½ inch NPS and larger is called *large* or *large bore.*

Some valves are made with no "ends." There are two different styles, wafer valves and lug valves, both of which are designed to be used with flanges. The wafer-style valve is used between mating flanges. Its circular body fits just inside the circular bolting pattern of the flanges. When tightened, the extended-length flange stud bolts cause the flanges to seal the ends of the valve and hold the valve in position. In this design, the valve does not contribute to holding the pipeline together. Wafer-style valves are shown in Figures 4-4 and 5-2. The lug-style valve is also circular but has projections (called *lugs*) with threaded holes spaced around its perimeter. The locations of the holes match those of the mating flanges, and the threads match the flange stud bolts used in assembly. The lug-style valve can be situated between flanges by using short stud bolts with each flange, with the valve body acting to hold the pipeline together, or the lug valve can be bolted to a single flange at the end of a pipeline. A lug-style valve is shown in Figure 5-1.

2

Gate Valves

The flow control element of a gate valve (called a *gate, wedge,* or *slide*) enters the fluid path from the side and traverses it until the fluid path is completely closed off, stopping the flow. When the valve is open, the gate is entirely out of the fluid path. Thus flow is in a straight line, with very little resistance from the valve. Because the gate valve is symmetrical, either end can be the inlet, and thus flow can be from either direction through the valve. The form of control for which gate valves are suited is starting and stopping flow. Gate valves, as are other valve types used for this kind of control, are frequently referred to *stop valves* or *block valves*.

GATE VALVE DESIGN

Figure 2-1 shows a typical gate valve. A manufacturer might describe this valve as a "flanged end, bolted-bonnet, outside-screw-and-yoke, flex-wedge, gate valve." Each term describes the particular design features of the valve with respect to the end connections, body–bonnet joint, stem design, and gate design, in that order. Considering the choices for each of these design features in turn, the many variations of gate valve designs that are available will become evident. End connections have been discussed in Chapter 1 and, therefore, will not be repeated here.

Gate Valves

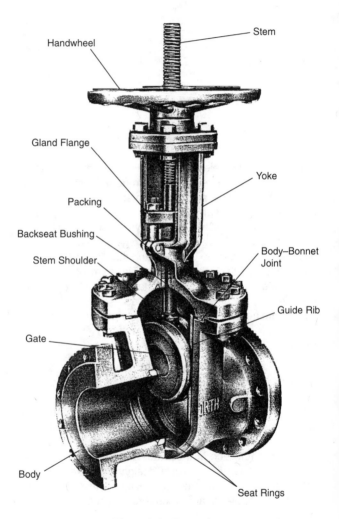

Figure 2-1. Gate valve

Gate Valves

1. Body–Bonnet Joint. The available designs of body–bonnet joints are the screwed, union, bolted-bonnet, pressure-seal, and welded-bonnet joints.

- The screwed joint is shown in Figure 2-2. The male thread on the bottom of the bonnet screws into the female thread in the top of the body. It is the simplest, least-expensive joint and is used

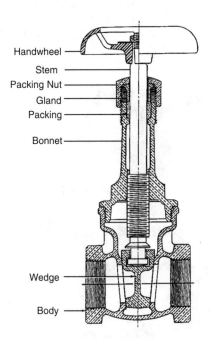

Figure 2-2. Screwed body–bonnet joint, inside screw rising stem (ISRS) design

on small bronze valves in which disassembly is seldom required.
- The union joint is shown in Figure 2-3. As in a pipe union, a female-threaded nut locks the unthreaded bonnet against the male-threaded body. It has an advantage over the screwed joint because the body and bonnet sealing surfaces do not rub together and wear when the joint is being tightened. This makes it preferable to the screwed joint when disassembly is necessary. The union nut also adds strength and stiffness to the joint. Because of the high torque required to

Figure 2-3. Union body–bonnet joint

assemble or remove the union nut, the union joint is limited to use on small valves.
- The bolted-bonnet joint shown in Figure 2-1 has mating flanges on the body and bonnet that are held together with studs or bolts and nuts. On small valves, it is common to have the bonnet held on by bolts that screw directly into threaded holes in the top of the body. A gasket is placed between the flanges to prevent leakage. Depending on the pressure capability of the valve, the gasket can be flat, solid material; filled spiral-wound material; or a metal ring. Flat gaskets are found on low-pressure valves; whereas spiral-wound gaskets and metal rings are used on medium- and high-pressure valves. Bolted-bonnet joints can be readily disassembled for repair and are used on all sizes of cast iron and steel valves.
- The pressure-seal joint is shown in Figure 2-4. In this design, internal pressure against the bottom of the bonnet causes it to press against a relatively soft seal ring. Because of its wedge-shaped cross-section, the seal ring deforms to press tightly against both the outside surface of the bonnet and the bore of the body, sealing the joint. The segmented thrust ring, which is held in a circular groove at the top of the body, absorbs all the thrust applied by the internal pressure. The hardened spacer ring between the seal ring and the thrust ring prevents deformation of the top of the seal ring. The retaining studs of the bonnet hold it in place and apply a preload to the seal ring before the valve is put into service. Pressure-seal joints are used on steel valves for high and very high pressure service.

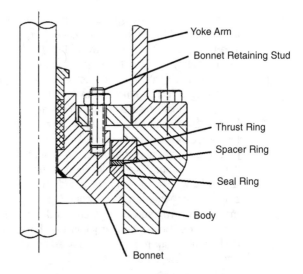

Figure 2-4. Pressure-seal body–bonnet joint

- A welded joint is shown in Figure 2-5. Welding the bonnet to the body ensures against leakage at the joint but makes disassembly of the valve much more difficult. The welded bonnet produces a lighter valve than does the bolted bonnet or the pressure seal. This joint is usually found only on small steel valves. Because repair of small steel valves is considered uneconomical, the difficulty of disassembly is not a disadvantage.

2. **Stem Design.** Gate valves are available with three stem designs: inside screw rising stem (ISRS), non-rising stem (NRS), and outside screw and yoke (OS&Y).

Gate Valves

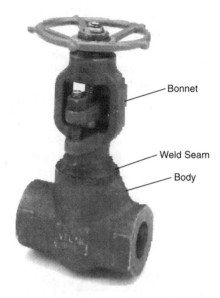

Figure 2-5. Welded body–bonnet joint

- The inside screw rising stem (ISRS) is shown in Figure 2-2. The right-hand thread on the stem mates with an internal thread in the bonnet so that by turning the valve handwheel clockwise the stem and gate are translated downward, closing the valve. (Turning the valve handwheel clockwise to close the valve is the accepted convention in the valve industry.) The height of the stem outside the valve indicates whether the valve is open or closed. The ISRS usually is used only on bronze valves, for which it is the standard.

Gate Valves

- A non-rising stem (NRS) is shown in Figure 2-6. This design is used where insufficient space above the valve limits stem movement upward. In this case a left-hand thread on the stem mates with an internal thread in the gate. An integral stem collar held in the bonnet keeps the stem from moving up or down but permits it to turn. A left-hand thread is used so that turning the handwheel clockwise closes the valve. Because the stem does not move upward, it can-

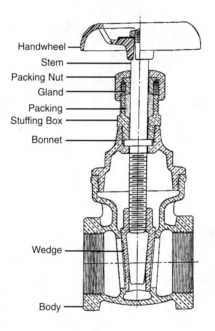

Figure 2-6. Non-rising stem (NRS)

not be used to determine whether the valve is open or closed. The NRS usually is found only on bronze and cast iron valves but is also available from some manufacturers on large steel gate valves. Using the NRS with a cast iron or steel gate valve requires that a non-galling nut (usually bronze) be placed in the gate to prevent excessive wear between the stem and the gate.

- The major disadvantage of both the ISRS and NRS designs is that the fluid flowing through the valve "wets" the stem threads, making them subject to corrosion and erosion and thus possible failure. However, there are environments—such as on off-shore oil-drilling platforms—in which external corrosion is a factor because of the severe conditions. In these exceptional instances, ISRS and NRS stems are preferred because the external corrosion outweighs the internal corrosion caused by the fluid passing through the valve. The outside-screw-and-yoke (OS&Y) design shown in Figures 2-1 and 2-7 prevents thread wetting because the threaded portion of the stem is outside the valve. The stem passes through a yoke, which is formed by two arms that extend up from the bonnet and joint to form a housing for a yoke nut (also called a *stem nut, yoke bushing,* or *stem bushing*). Unlike the other stem designs, the handwheel is attached to the yoke nut rather than to the stem. The yoke nut is free to turn but is held in place at the top of the yoke by a yoke-nut retainer, which also acts as a bushing for the nut. The yoke nut has a left-hand internal thread that mates with the thread on the

Gate Valves

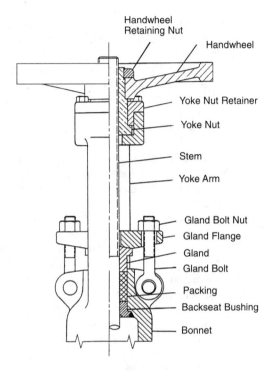

Figure 2-7. Section through yoke nut and stuffing box (packing chamber)

stem. The stem thread is left-handed so that turning the handwheel clockwise moves the stem downward, closing the valve. Being outside the valve, the threads can be lubricated easily when necessary. A grease fitting in the yoke-nut housing is usually provided for this purpose. On

Gate Valves

large valves, the yoke nut is usually held in ball or roller bearings to reduce friction with the yoke-nut retainer and to reduce handwheel force.

In this design the stem does not turn and may be pinned to the gate; however, it usually has a T head that fits into a slot in top of the gate, allowing the gate to freely align itself with the seats. The position of the gate within the valve is easily determined from the position of the top of the stem. The OS&Y design is standard for all sizes of cast iron and steel gate valves.

In all three stem designs, external leakage along the stem is prevented by the use of deformable non-metallic packing. The packing is located in the bonnet in a circular cavity commonly known as the *stuffing box* or *packing chamber.* In valves using ISRS and NRS stem designs, the packing is held in place by a tubular gland backed by a hollow packing nut that threads onto the top of the bonnet, as shown in Figures 2-2 and 2-6. In valves using the OS&Y stem design, the packing arrangement is more complicated and is shown in Figure 2-7. The packing is held in place by a gland, which is backed by a gland flange. The gland flange is restrained by a pair of gland bolts and nuts, which are, in turn, fastened to the bonnet using eyes and pins or some comparable arrangement. Both methods provide a means for adjusting the packing in the event leakage occurs. Tightening the packing nut or gland nuts compresses the packing, causing it to deform and press tighter against the stem.

Both ISRS and OS&Y stem designs usually provide for changing the packing without removing the valve from the pipeline by having a shoulder machined on the stem near the gate end. When the valve is completely open, this shoulder bears against a machined *backseat* in the bonnet forming a fluid-tight seal. The packing can then be removed without leakage. The backseat can be integral with the bonnet, or it can be a separate bushing that is threaded, pressed, or welded into the bonnet. A stem shoulder and a backseat busing are shown in Figure 2-1.

3. Gate Design. The most common gate designs available in the gate valve are the wedge gates, so called because in cross-section they are shaped like wedges. They are used in conjunction with two body seats set at angles of from 3° to 6° to the stem centerline. The wedging action obtained when closing the valve forces the gate tightly against both seats, producing two sealing surfaces and making the valve tight against flow in either direction. Three styles of wedge gates are avail-

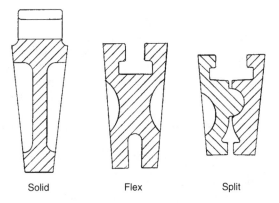

Figure 2-8. Wedge gate designs

Gate Valves

able: the one-piece solid (commonly known as a *solid wedge*), the one-piece flexible (commonly known as a *flex wedge*), and the two-piece *split wedge*. These are shown in Figure 2-8.

- The solid wedge gate was, at one time, the most frequently used design. It is still the simplest, most economical style and is the most resistant to corrosion and vibration. It is also ideal for steam service and turbulent flow. A solid wedge gate, modified by adding a stem cavity, is used with the NRS. It is the standard on bronze, cast iron, and small steel valves. The drawback of the solid wedge gate is its rigidity, which does not allow it to accommodate seat distortion and makes it prone to sticking when subjected to temperature extremes. In addition, sealing depends on precise machining and fitting of the wedge and body seating surfaces. The larger the valve, the more difficult it is to obtain a good fit with the solid wedge gate. Consequently, it is not readily available in large steel valves.
- The flex wedge gate is cast with a circumferential groove around its perimeter or one is machined into it. This groove produces the flexibility that allows the seating surfaces of the wedge to move independently and adapt to minor inaccuracies in seating-surface angles and movement of the valve seats owing to pipeline loads or thermal expansion of the piping system. It minimizes wedge sticking when the valve is closed when hot and opened when cold. This ability to flex while retaining one-piece construction makes the flex-wedge design the best one for most applications. It is now the standard for large

steel valves. It is not used on bronze or cast iron discs, however, because neither metal possesses the combination of strength and flexibility required by the flex-wedge design.
- The split wedge gate is made in two parts with a ball-and-socket type joint between them, thereby providing complete flexibility to compensate for seat movement and seat angle machining tolerances. The split wedge gate is found on small bronze valves in which the flex wedge is not practical and on large valves made of stainless steel and other expensive materials. In the case of expensive materials, it is more economical to provide split wedge gates than the thick body walls needed to resist seat movement owing to line loads and thermal expansion. The split wedge gate is not recommended for high velocity or turbulent flow in which the two halves could vibrate and chatter.

Wedge gates are typically guided by ribs located either in the valve body or on the wedge, and fitting into mating slots in the sides of the wedge or in the valve body, respectively. This arrangement prevents the force of the flowing fluid from pushing the unseated gate against the downstream seat when the valve is being opened or closed, causing wear of the seating surfaces.

In addition to the wedge gates, another design, the double disc, is available. It is used with body seats whose faces are parallel to the stem centerline. It has two discs and an internal wedge-shaped spreading device that forces them outward against the seats after the gate is in the closed position. When opening, the force of the discs on the seats is relieved before the gate is raised, avoiding friction and wear of the seating surfaces. The double-disc design is used where high reliability of the seating surfaces is required, such as when the valve is inaccessible for repair. It is widely used in water works and in oil and gas cross-country pipelines. As is the split wedge, the double-disc gate is unsuited for steam and turbulent flow. Because of their greater complexity, gates of this design are more costly than are wedge gates.

All gate designs are used with body seats that are either integral with the body (machined surfaces on the body), separate seat rings pressed or screwed into the body, or hard material weld overlayed in the valve body and then machined as an integral seat. For steel valves, which are used in high-temperature and high-pressure applications, the separate seat rings are also seal-welded to the body. This ensures against leakage between the body and the seat ring.

GATE VALVE APPLICATIONS

The gate valve is the most frequently used valve in industrial and power piping applications, because the most common valve application is starting and stopping flow. It is the best choice for a stop valve in most cases, because it usually has one or more of the following advantages over other stop valves:

1. It offers little resistance to flow when it is open. When closed, the two seats reduce the chance of internal leakage.
2. It can be used in more places because it is manufactured in a wider range of sizes, pressure classes, and materials than any other type of stop valve. Sizes of

cast and forged body gate valves range from ¼ inch to 30 inches. Larger sizes made of fabrications are available and usually are custom-made by special order. Gate valves can handle fluid pressures from vacuum to over 4,500 psi, and they are available in all types of valve materials: bronze, cast iron, steel, and corrosion-resistant alloys.

3. It can be used with all clean fluids: liquid, gas, and vapor. Fluid temperatures are limited only by the materials of construction.

4. It is manufactured by many companies and in large quantities because of its popularity. Consequently, it is inexpensive. The gate valve is typically the least-expensive valve in a given size.

5. It is versatile. The design options described previously can be combined in different ways to meet many specific applications. Some typical examples follow:

    - For low-temperature, low-pressure water service, a bronze gate valve with a union bonnet, an ISRS, and a solid wedge gate would be suitable for small lines. For large lines, a cast iron valve with a bolted-bonnet joint, an OS&Y, and a solid wedge gate would be the choice.
    - For 1,000 psi saturated steam (about 550°F) service, a carbon steel gate valve with a pressure-seal or bolted bonnet, an OS&Y stem, and a solid or flex wedge gate would be right. For small lines, the valve would have a welded bonnet and a solid wedge gate.
    - For 10% hydrochloric acid at ambient temperature service, the valve would be stainless steel with a bolted bonnet, an OS&Y stem, and a split wedge gate. On a small valve, the wedge may be solid.

As good as it is, the gate valve does have shortcomings. For a given size, it is usually larger and heavier than other types of stop valves. Because of the long distance the gate must travel from the open to the closed position, the gate valve is less desirable than other stop valves when frequent manual operation is required. In addition, gate valves should not be used with fluids containing solids (e.g., "dirty" fluids and slurries), which can build up in the cavity between the seat rings or between the halves of a split-wedge or double-disc gate. Such build-up can prevent proper seating of the gate.

The gate valve should not be used for regulating flow (called *throttling*). Because a high percentage of total flow through a gate valve occurs when it is only 5% to 10% open, throttling could be done with the wedge only slightly open, resulting in wedge vibration and erosion of the seats owing to high fluid velocity. In addition, looseness at the wedge guides could cause the seating surfaces to drag against one another when the wedge is only slightly open; therefore, frequent operation near closure could cause extensive wear on the downstream seats.

KNIFE GATE VALVE

A style of gate valve different from the standard gate, the knife gate valve (also called a *slide valve*), is shown in Figure 2-9. As can be seen, it is a much simpler design.

The knife gate valve has no bonnet, eliminating the body–bonnet joint. External leakage is prevented by packing placed in a recess that goes completely around the inside of the top of the body and around the gate. Because the gate extends above the top of the body even when the valve is closed, the packing seal is never broken. The packing is held in place by a one-piece flanged packing gland that can be adjusted to maintain

Gate Valves

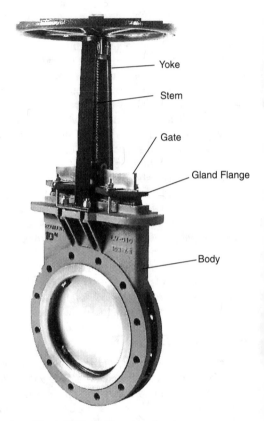

Figure 2-9. Knife gate valve

the seal. The stem is an OS&Y design, with the yoke attached directly to the top of the body. The stem is completely outside the body and attaches to the gate with a clevis and pin.

Gate Valves

The gate itself is simply a flat plate that is used in conjunction with two different body seat designs. These are shown in Figure 2-10. In one design, the edge of the gate is pressed against a single elastomer seat situated in a groove in the body. The elastomer seat extends around the bottom and up both edges of the gate, up to the packing recess. This design produces tight, bidirectional shut-off; however, use of elastomeric material limits the fluid temperature the valve can handle.

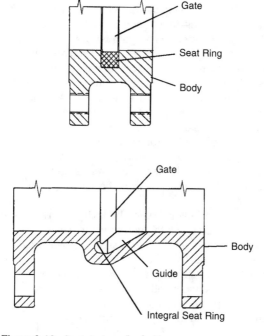

Figure 2-10. Seat designs for knife gate valves

In the other body seat design, a single seat is used. The seat is either integral with the body or is a separate ring, with its face parallel to the gate. The lower edge of the gate is beveled on the side away from the seat, so that just as the gate is seating, guides in the body force the gate against the body seat for tight shut-off. This design is most effective for fluid flow in the direction of the seat ring. Flow in the other direction tends to push the gate away from the body seat.

The bonnetless design of the knife gate valve imposes a limitation on its use in regard to fluid pressure. The large sealing area around the gate and its difficult configuration make it difficult to achieve a tight seal against external leakage; therefore, knife gate valves can only be used in low-pressure applications. Their simple design, however, facilitates knife gate valves being made as weldments, which allows them to be manufactured in much larger sizes than are standard gate valves. Also, for the same size valve, the knife gate is lighter, smaller from end-to-end, and usually no more expensive than the standard gate, especially in large sizes. Almost all knife gate valves are made with flanged ends or in the lug style from carbon or stainless steel. Standard sizes range from 2 inches to 36 inches; larger sizes must be custom-made by special order.

The major advantage of the knife gate valve is the ability of its gate to cut through solids that would fill the space between the seat rings in a standard gate valve and prevent closure. This makes it ideal for use with slurries, solids carried in air, and in dry service such as on storage hoppers. Consequently, knife gate valves are used extensively in process industries such as pulp and paper, waste treatment, and mining (ore processing).

3

Globe Valves

The body of a globe valve is configured so that the flow control element (called a *disc*) is moved along the axis of the fluid path. The disc is always in the fluid path. The valve opening is annular. The form of control for which globe valves are best suited is regulating flow volume (called *throttling*); however, they also can be used as stop valves (for starting and stopping flow).

GLOBE VALVE DESIGN

Figure 3-1 shows a typical globe valve. It can be described as a "flanged end, bolted-bonnet, outside-screw-and-yoke, spherical disc, globe valve." Each term describes the particular design features of the valve with respect to the end connections, body–bonnet joint, stem design, and disc design, in that order. The different designs with respect to end connections have been described in Chapter 1, and the body–bonnet joints used for globe valves are the same as those used for gate valves and have been described in Chapter 2. The different designs for stems and discs are discussed below:

1. Stem Design. Globe valves are available with two stem designs: the inside screw rising stem (ISRS) and the outside-screw-and-yoke (OS&Y) designs.

Globe Valves

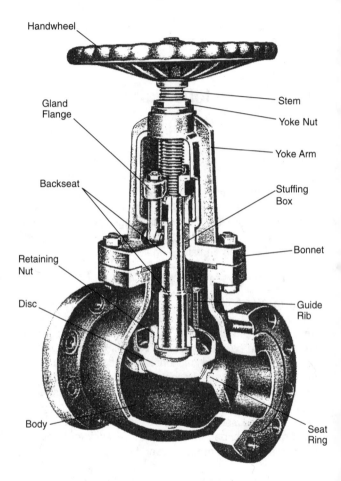

Figure 3-1. Globe valve

Globe Valves

- The inside screw rising stem (ISRS) design is identical to that used in the gate valve and is shown in Figure 2-2. A right-hand thread on the stem mates with an internal thread in the bonnet so that by turning the valve handwheel clockwise the stem and gate are translated downward, closing the valve. The ISRS usually is used only on bronze valves, for which it is the standard.
- The outside-screw-and-yoke (OS&Y) design shown in Figure 3-1 prevents thread corrosion, because the threaded portion of the stem is outside the valve shell. The stem passes through a yoke, which is formed by two arms that extend up from the bonnet and join to form a housing for a fixed yoke nut (also called a *stem nut, yoke bushing,* or *stem bushing*). As in the ISRS, the OS&Y stem has a right-hand thread so that turning the handwheel clockwise closes the valve. Being outside the valve, the threads can be lubricated easily when necessary. This design is standard for large cast iron and steel valves.

On large, high-pressure, cast steel globe valves, some manufacturers use the arrangement shown in Figure 2-7, in which the handwheel drives the yoke nut, and not the stem. The stem has a left-hand thread, so that turning the handwheel clockwise closes the valve. Also, the yoke nut is usually mounted in ball or roller bearings to reduce friction between the yoke nut and the yoke.

Prevention of external leakage along the stem of a globe valve (*stem packing*) is the same as that for the gate valve and has been described in Chapter 2. The use of a backseat to permit in line packing replacement has also been covered in Chapter 2.

Because the distance traveled by the disc on a globe valve is relatively short, the position of the disc is not readily evident from the stem position for either stem design. Consequently, most manufacturers provide disc-position indicators as a globe valve accessory.

2. Disc Design. Globe valve discs are not fixed on the end of the stem. In the most common design, the end of the stem has a small flange, which is inserted into a threaded cavity in the back of the disc. A disc retaining nut fits over the stem and screws into the disc cavity, locking the stem flange in the disc. This arrangement permits the disc to turn relative to the stem and not the body seat when seating the disc, thereby reducing wear and galling of the seating surfaces. Also, depending on the amount of looseness between the disc retaining nut and the stem, compensation for minor misalignment of seating surfaces is provided.

On some small valves the top of the disc is slotted, and the stem flange is slid into the slot from the side. The walls of the valve body adjacent to the disc constrain the disc laterally and keep it from sliding off the stem.

In small and low-pressure valves there usually is adequate stiffness in the stem-and-disc assembly to prevent excessive flow-induced vibration and chatter; however, in large and high-pressure valves some form of disc guidance is required. This usually is done by guiding the disc in a cylindrically bored body neck or cylindrical body insert. Alternatively, machined ribs in the body neck can be used. There are two different approaches to disc guiding: one is to have extensions from the bottom of the disc "pilot" in the bore of the seat ring, and the other is to have a pin extending from the disc bottom pass through a hole in a bridge across the seat ring.

Globe Valves

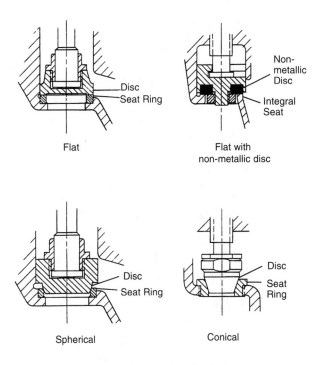

Figure 3-2. Globe valve disc designs

Several different disc seating-surface geometries are used: flat, spherical, and conical (also called *plug-type*). They are shown in Figure 3-2.

- The flat seat is easy to machine and lap to close tolerances, is not susceptible to misalignment, and is faster and easier to maintain with the valve in line. Because its contact area is wide,

large closing forces are required to achieve tight sealing. This seat is easily modified to use nonmetallic disc inserts, which are common on small, low-pressure valves. These disc inserts make dependable, tight seats when used with air and other gases and are suitable for most other low-temperature fluid services. However, small particles of foreign matter can become imbedded in the relatively soft disc insert and can prevent tight seating.
- The spherical seat is more difficult to machine and is susceptible to misalignment; however, it provides virtual line contact and lower closing force for tight seating. The narrow seating area on the disc is subject to erosion in high-velocity flow.
- The conical seat is easier to machine than is the spherical seat but also is susceptible to misalignment. The wide contact area requires a high closing force to seal the conical seat; however, this makes it less susceptible to leakage from nicks and cuts caused by foreign matter.

All three seat geometries are used with body seats that are either integral with the body (machined surfaces on the body), separate seat rings pressed or screwed into the body, or hard material weld overlayed in the valve body and then machined as an integral seat. For steel valves the separate seat rings can be seal-welded to the body, ensuring against leakage between the body and the seat ring.

The internal configuration of a globe valve causes the fluid to go through two 90° changes of direction, resulting in higher resistance to fluid flow and greater disturbance than would occur in valves with straight-through flow.

Globe Valves

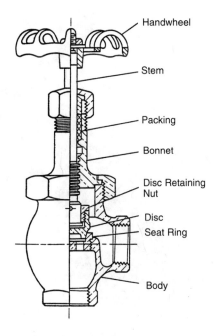

Figure 3-3. Angle valve

Alternatives to the standard globe valve configuration that have lower resistance to flow and reduced disturbance are available. One is the angle valve shown in Figure 3-3. Here, the fluid must make only one 90° change of direction. Of course, the angle valve is installed where the pipeline makes a 90° bend. Although this is not always possible, using the angle valve instead of a standard globe valve where possible eliminates a 90° pipe elbow. Another alternative to the standard globe valve is

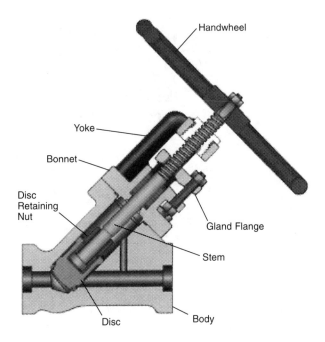

Figure 3-4. Y-pattern globe valve

the Y-pattern globe valve shown in Figure 3-4. Resistance to fluid flow is much reduced and the location of this valve in a pipeline is not limited. The Y-pattern globe is widely used for boiler blow-off applications.

GLOBE VALVE APPLICATIONS

The standard globe valve is the most common valve encountered in industrial and power piping flow regulation applications. This is because it usually has one or more of the following advantages over other valves used in throttling service:

1. The variation of fluid flow with respect to disc position is approximately linear in the globe valve. That is, the flow rate is directly proportional to the distance the disc travels. For instance, opening the valve half-way results in a flow rate of approximately 50% of the flow rate of the fully open valve. This linear relationship makes it possible to predict the flow rate from the position of the handwheel (and the disc), and it makes the globe valve a preferred valve for throttling service. The relationship between the position of the flow control element and the flow rate is known as the *valve flow characteristic.* Valve flow characteristics are discussed in greater detail in Chapter 11.
2. It is made in a wider range of sizes, pressure classes, and materials than other types of throttling valves. Sizes range from ⅛ inch to 16 inches. Globe valves can handle fluid pressures from vacuum to over 4,500 psi. They are available in all types of valve materials: bronze, cast iron, steel, and corrosion-resistant alloys.
3. It can be used with all clean fluids: liquid, gas, and vapor. Fluid temperatures are limited only by the materials of construction.
4. Its seating surface can be serviced more easily in line than can other throttling valves because of the position of its body seating surface. The body seat is completely visible when the valve bonnet is removed.

The standard globe valve does have some drawbacks. It is usually larger and heavier than are other types of throttling valves, especially in large sizes. In addition, it is not suitable for use with fluids containing solids (e.g., "dirty"

Globe Valves

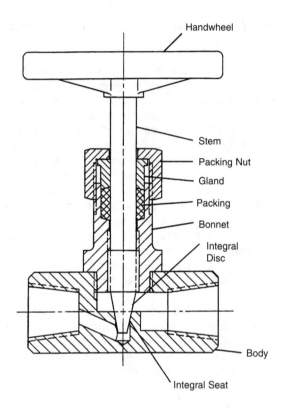

Figure 3-5. Needle valve

fluids and slurries) that can build up in the body where fluid direction changes and behind the disc, eventually clogging the valve.

Although primarily used for regulating flow volume, the globe valve can be used as a stop valve. Its short stroke and tight shut-off make it useful for frequent operation applications in which its high flow resistance is acceptable.

It is recommended that globe valves be installed so that the fluid pressure is under the disc. Globe valve bodies are commonly marked with flow-direction arrows to facilitate this installation. When a globe valve is installed so that the fluid pressure is on top of the disc, the resistance to fluid flow is not substantially affected. However, when the disc is unseated, the full force of the fluid on the back of the disc is taken by the small flange on the end of the stem, and the stem packing is subjected to the high inlet pressure rather than the lower discharge pressure.

NEEDLE VALVE

A simple form of globe valve, the needle valve, is shown in Figure 3-5. It does not have a separate disc or seat ring. The end of the stem is machined almost to a point to form an integral conical disc that seats in a mating conical hole in the body. The name *needle valve* derives from the needle-shaped stem.

Needle valves are primarily made of bronze, carbon steel bar stock, and stainless steel bar stock in sizes from $\frac{1}{8}$ inch to 1 inch. The bronze valves can be used at pressures up to about 400 psi; the steel valves are suitable for either liquid or gas service for pressures up to 5,000 psi. Needle valves are available in both standard and angle configurations. They are used with pressure-level and liquid-level gauges, orifice plates, instrument lines, and other applications in which fine regulation of flow is required. Needle valves are particularly well suited for the tight spaces behind hydraulic control panels.

4

Check Valves

Check valves come in two basic styles. In the first, the flow control element, called a *disc, flapper,* or *plate,* rotates about an axis perpendicular to the fluid path. In the second style, the flow control element, which can be a disc, piston, or ball, moves along the axis of the fluid path. In both styles the force of the fluid causes the flow control element to unseat automatically, opening and maintaining the fluid path through the valve. If the flow stops, the weight of the flow control element, an auxiliary spring, or both, causes the flow control element to return to its seated, closed position. In the case of prevented reverse flow, the back-pressure of the fluid assists in seating the flow control element and makes for a tighter seal.

The form of fluid control of the check valve is the prevention of fluid flow reversal. Two typical reverse-flow prevention applications are at pump discharges and at places where different portions of a piping system join a common header. Check valves usually are installed at the outlet of a centrifugal pump to prevent back-flow of fluid from a higher elevation when the pump is not pumping. Back-flow causes the pump impeller to turn backward at a higher than normal speed, with possible damage to the pump. Back-flow may even cause downstream tanks or processing vessels to drain. Check valves are used to isolate separate services connected to a common header to prevent fluid flow from one service to another, thereby preventing contamination. Check valves are also used to isolate low-pressure-rated systems and equipment from downstream high-pressure surges.

Other types of valves have variations in the design of their components, but their basic configurations are fixed. In contrast, check valves have different configurations. The most common are the swing check valve, tilting-disc check valve, wafer check valve, lift check valve, and stop-check valve. Each is discussed below.

SWING CHECK VALVE

Figure 4-1 shows a swing check valve. It is the most popular style of the several designs of the rotating flow control element check valve. The valve in Figure 4-1 is a "flanged end, bolted-cap, swing check valve." The different designs with respect to end connections have been discussed in Chapter 1. The body–cap joint of the swing check valve is comparable to the body–bonnet joint of the gate valve, and its different designs have been discussed in Chapter 2.

In the swing check valve design in Figure 4-1, a short integral stud extending from the back of the disc passes through a hole in a link, called a *hinge,* and is held in the link by a retaining nut. The disc is not held tightly against the hinge but is free to rotate, which helps distribute wear and erosion evenly on the disc seating surface. The disc retaining nut is locked in place with a cotter pin or some other locking device. If it is not locked in place, disc rotation can cause it to unscrew from the stud, and the disc can fall out of the hinge. The hinge is held at its other end by a pin, which is fixed in the valve. The body seat can be integral or a separate ring, as shown.

Check Valves

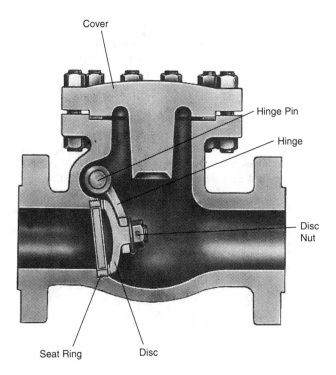

Figure 4-1. Swing check valve

In operation, the force of a fluid entering the valve from the inlet overcomes the weight of the disc and hinge and causes them to rotate around the hinge pin and swing upward, opening the valve. If the fluid stops, the weight of the disc and hinge causes the disc to swing down to its seated position. If fluid enters from the outlet end, it presses the disc against the seat ring and the valve stays closed.

In Figure 4-1, the hinge pin is located in the neck of the body, above the seat ring. Another less common design is one in which the hinge pin is suspended from the underside of the cap. The in-body design provides assurance of proper alignment and location of the disc and permits modification; whereas cap-mounting facilitates valve assembly and disassembly. Body mounting itself has two designs. In the first, the hinge pin is fixed in a bracket that is bolted to the inside of the body neck. In the second, the pin passes through holes in the sidewalls of the body neck. The holes are threaded, and plugs are inserted to retain the hinge pin and seal the body. The first design eliminates the possibility of external leakage through hinge pin holes. The second design uses fewer parts and can be modified to add an outside lever-and-weight assembly. The lever-and-weight assembly is shown in Figure 4-2. The hinge pin is extended out through one side of the body. A small stuffing-box–packing–packing-nut assembly, such as a stem seal (as shown in Figures 2-2 and 2-6), is used at the extended hinge pin to prevent external leakage. The lever is attached to the extended pin. The weight can be positioned on the lever, as shown in Figure 4-2, and added to the disc weight to either provide quicker closing or prevent opening until a desired fluid force is attained. The weight can also be positioned at the other end of the lever to act as a disc counterweight so the valve will open with low fluid force.

A slightly different configuration of the swing check valve found on small bronze valves is the "regrinding" type. The body has a Y-pattern, with both the seating surface and the valve cap tilted at 45° to the flow path. (This is similar to the Y-pattern globe valve shown in Figure 3-4 and the ball check valve shown in Figure 4-7.) When the cap and the disc are removed, the body seat is fully exposed and eas-

Check Valves

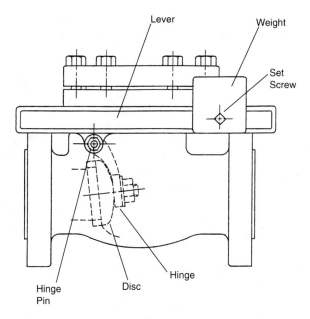

Figure 4-2. Swing check valve with lever and weight

ily refinished (or *reground*) without removing the valve from the line.

As noted previously, the swing check is the most popular check valve. It is a simple design and has a straight-through flow path, with low resistance to flow when open. It is suitable for all clean fluids and is made in a wider range of sizes, pressure classes, and materials than any other check valve configuration. Swing check valve sizes range from ¼ inch to 36 inches. These valves can be made to handle fluid pressures in excess of 4,500 psi, and they are available in all types of valve materials:

bronze, cast iron, steel, and corrosion-resistant alloys. The swing check valve is the standard for cast iron and cast steel valves.

The swing check valve does have some drawbacks. It should not be used for high-velocity flow or in applications in which there are frequent flow reversals, which can produce disc instability and result in accelerated hinge and hinge-pin wear. In addition, it should not be used with fluids such as slurries that can cause build-up on seatings surfaces, preventing tight closure. A major drawback of the swing check valve is its tendency to slam closed, producing high-impact stresses on the seating surfaces and noise and vibration (called *water hammer*). Slamming is the result of the disc not completely closing before full flow reversal. This is due to the long distance the disc must travel from the open to the closed position.

TILTING-DISC CHECK VALVE

Another design of a rotating element check valve is the tilting-disc check valve shown in Figure 4-3. The valve in Figure 4-3 is a flanged end, split body valve. Tilting-disc check valves are also available with butt-weld ends, and with a one-piece body and a cap, as in the swing check valve. Body–cap joints can be bolted or pressure-seal caps. In Figure 4-3 the two body halves are bolted together, capturing the seat ring between them. In other designs the seat ring either is integral with the inlet body half or is a separate ring that is screwed or welded in the body half. The disc pivots on pins that pass through holes in the outlet body half. The holes are covered with small caps to prevent external leakage. The pivot pins are located close to but above the center of the disc. Owing to pivot-point offset, the fluid force on the disc is unbalanced. As a result,

Check Valves

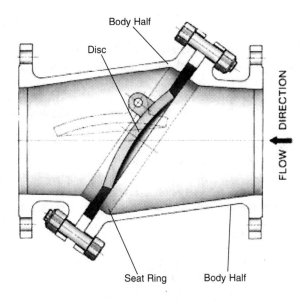

Figure 4-3. Tilting-disc check valve

the disc unseats and seats, just as the swing check valve does, but with a substantially shorter distance to travel from the open to the closed position.

The tilting-disc check valve is better-suited for frequent flow reversals than is the swing check valve and is less prone to slamming because of the shorter travel of the disc. Tilting-disc check valves are available in steel and corrosion-resistant alloys, in sizes from 2½ inches to 24 inches, and they can be made to handle pressures in excess of 4,500 psi.

WAFER CHECK VALVE

A third design of a rotating flow control element check valve is the wafer check valve shown in Figure 4-4. Although called a wafer check, this valve design is also available with a lug-type body and with flanged or butt-weld ends. It features two half-discs, called *plates*, that pivot on a center post, called the *hinge pin*. The plates are held in the closed position against integral body seats by torsion springs also mounted on the hinge pin and restrained from free turning by a pin situated parallel to the hinge pin. In Figure 4-4 the pins penetrate body, but designs without body penetration are also available. Fluid force on the

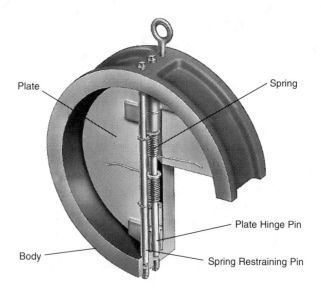

Figure 4-4. Wafer check valve

plates is transmitted to the springs, causing them to "wind up" and allow the plates to unseat. The degree of plate opening is proportional to the fluid force and spring strength. Faster plate closing can be obtained by using stronger springs; however, stronger springs require higher fluid force on the plates to open the valve.

The wafer check valve offers several advantages over swing check and tilting-disc check valves:

1. It is smaller and lighter than the same size of swing check or tilting-disc check valve.
2. Its short travel and quick plate closing reduce slamming and make the wafer check valve more suitable for applications in which there is frequent flow reversal.
3. It is available in much larger sizes—up to 72 inches. Wafer check valves are available in all types of valve materials: bronze, cast iron, steel, and corrosion-resistant alloys. They can be made to handle fluid pressures of over 4,500 psi.
4. It easily can be lined with non-metallic material for use with corrosive fluids or fluids containing abrasive particles.

The wafer check valve also has drawbacks. As in the tilting-disc check valve, it has higher resistance to flow than does the swing check valve because the plates are always in the flow path. Unlike the swing check and tilting-disc check valves, which need only gravity to operate, the wafer check is solely dependent on the springs. If the spring should fail from fatigue, corrosion, or both, the valve may be rendered inoperative.

LIFT CHECK VALVE

The lift check is the other style of check valve in which the flow control element—either a piston, disc, or ball—

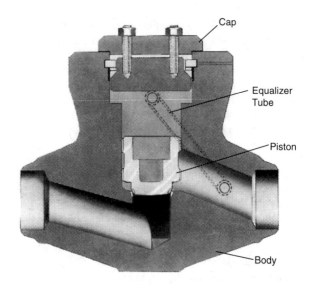

Figure 4-5. Lift check valve

moves along the fluid path. Figures 4-5, 4-6, and 4-7 show lift check valves with the different forms of flow control elements. The valve in Figure 4-5 is a "butt-weld pressure-seal cap, piston check valve." It is called a piston check valve because of the shape of its flow control element (called a *piston*). The valve in Figure 4-6 is a "threaded end, threaded cap, lift check valve," and the one in Figure 4-7 is a "socket-weld end, bolted-cap, ball check valve." In the ball check valve the flow control element is a solid ball. All the different designs of valve ends and body–cap joints available for lift check valves are the same as those available for swing check valves.

In operation, the force of a fluid entering the valve from

the inlet overcomes the weight of the flow control element and lifts it, opening the valve. If the fluid stops, the weight of the flow control element causes it to drop down to its seated position. If fluid enters from the outlet end, it presses the flow control element against the seat, and the valve stays closed.

The piston and disc flow control elements must be guided to ensure alignment of the seating surfaces. This is done with the piston by cylindrically boring the neck of the body to provide a guiding surface. For the disc, it is done by providing disc extensions that go down inside the seat ring, by a disc guide integral with the cap, or by a separate

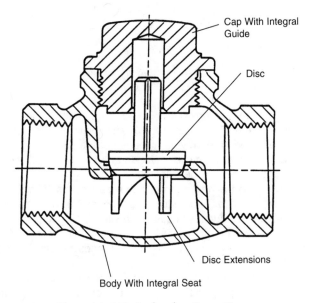

Figure 4-6. Lift check valve disc guides

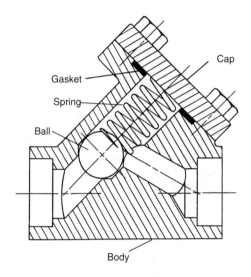

Figure 4-7. Ball check valve

disc guide. Disc extensions and a cap with an integral guide are shown in Figure 4-6. The relatively tight clearances needed to provide guidance of the piston and disc make them susceptible to sticking if used with fluids containing suspended particles.

Because the ball in the ball check valve is self-aligning, close guidance is not required for tight seating. As a result, this valve is much less susceptible to sticking and is suitable for use on dirty fluids. Wear on the ball is evenly distributed over its surface because it tends to rotate when unseated.

The configuration of the lift check valve makes it easy to add a spring above the flow control element, as was done on the ball check valve in Figure 4-7. The added spring causes faster closing and tighter seating of the flow control element. In addition, large piston check valves can be provided with equalizer tubes, as shown in Figure 4-5.

The equalizer tube connects the cavity above the disc with the downstream side of the valve, assisting the cavity above the piston in draining, ensuring rapid and full opening. Under conditions of back-flow it allows high-pressure fluid above the piston and resultant tighter seating.

The lift check valve is a simple, rugged design with no moving parts other than the flow control element, making it very reliable. The flow control element is easily replaced and the seating surface made accessible for repair by removing the valve cap. Flow control element travel is short, making the lift check valve resistant to slamming and applicable in services in which there are frequent changes of flow direction.

Lift check valves are suitable for all clean fluids, and the ball check valve can also be used with dirty and viscous fluids. Both are made in a wide range of sizes, pressure classes, and materials. Lift check valve sizes range from ¼ inch to 24 inches. However, the ball check valve is made only in small sizes—up to 2 inches. Both can be made to handle fluid pressures in excess of 4,500 psi, and both are made of bronze, steel, and corrosion-resistant alloys. They are not generally available in cast iron. The lift check valve is the standard for small, forged steel valves.

A disadvantage of the lift check valve is the higher pressure drop, resulting from the flow control element always being in the fluid path. Another disadvantage of lift check valves other than the ball check valve is that they cannot be used with dirty or gum-forming fluids that will cause the piston or disc to stick.

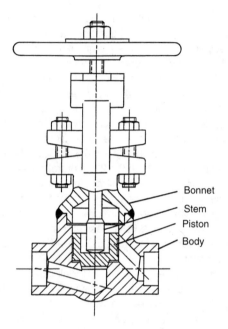

Figure 4-8. Stop-check valve

STOP-CHECK VALVE

Replacing the cap of a lift piston check valve with a bonnet and adding a stem, but not connecting it to the disc, produces a hybrid known as a stop-check valve (also called a *non-return valve*). Figure 4-8 shows a "socket-weld end, welded-bonnet, stop-check valve." When the stem is in the open position, the valve functions as a lift check valve. When the stem is closed the stop-check valve functions as a stop valve, preventing flow in either direction. In addition, when closed, the stem adds force to the piston to enhance tightness under reverse-flow conditions.

When a check valve does not seat reliably enough to handle high-pressure back-flow in critical isolation applications, a stop valve is required. The stop-check valve makes it unnecessary to provide a separate stop valve. For example, when more than one boiler is connected to a main steam header, a stop-check valve is installed in the pipeline between each boiler and the header. When the stem is in the open position, the valve acts as a check valve and prevents flow of steam from the header back to a boiler during boiler start-up or shut-down. When the stem is closed, tight sealing is assured, making it safe to work on the isolated boiler at the same time the other boilers remain in service.

Stop-check valves are available in cast iron, steel, and corrosion-resistant alloys. They are made in sizes from 2½ inches to 14 inches and can handle fluid pressures in excess of 4,500 psi.

CHECK VALVE INSTALLATION AND SIZING

Unlike other valve types, which can be installed in any position with little or no effect on their operation, most check valves have installation limitations. The swing check, tilting-disc check, and lift check valves all depend on gravity for the reseating of their flow control element; therefore, they must be installed and oriented so that gravity can have its desired effect. Obviously, all check valves must be installed so that flow is permitted in the proper direction, and they are usually marked with flow-direction arrows to facilitate correct installation. For other installation considerations the following rules apply:

1. All check valves can be used in horizontal pipelines.
2. Swing check valves and tilting-disc check valves can be placed in vertical lines and at any angle between horizontal and vertical, as long as the direction of fluid flow is upward. If the valve is mounted for downward flow, the disc will hang in the open position. Even with reverse flow, seating is not ensured.
3. The lift check valve can be installed at angle of up to 45° from horizontal. At larger angles the friction on the side of the disc and piston will retard closing. Fluid flow can be in either an upward or downward direction.
4. For other than vertical installations, the swing check, tilting-disc check, and lift check valves must be installed so that the top of the valve is above the pipeline axis, preferably in the vertical plane. The valve can be rotated around the pipeline axis so that the cap is not in the vertical plane. However, the amount of rotation should be limited to about 30° so that friction on the side of the flow control element or pivot points does not retard closing.

Because the wafer check valve and the spring-loaded lift check valve do not depend on gravity for reseating, they may be installed in any position. However, the spring-loaded lift check valve must be oriented so that the top of the valve cannot be rotated more than 90° from vertical, because if it were then it would be possible for small particles of dirt and scale to collect and jam open the piston or disc.

Another important difference between check valves and other valve types is that the amount of opening of a check valve is dependent on the force of the fluid passing through it, and not on the input of the valve operator. However, check valves are designed to function in the fully open position. When fully open, a check valve has less resistance to flow, and its flow control element is more stable. Consequently, it is important that check valves be properly sized to be completely open under normal flow conditions.

To aid in determining correct check valve size, manufacturers have developed different approaches. One manufacturer supplies formulas for different designs (e.g., swing check and tilting-disc check), which are based on the specific volume of the fluid, to calculate the fluid velocity required to fully open the valves they manufacture. The maximum valve size needed to obtain the required velocity can then be determined from the fluid flow rate the valve is to handle. Another manufacturer supplies a sizing parameter for each of the different designs and sizes of their check valves. This unique sizing parameter is compared with a value calculated from the flow rate and density of the fluid. If the sizing parameter is less than the calculated value, the valve will be fully open.

Manufacturers' sizing approaches are the result of testing on their own valves and, therefore, should be used only with these valves. When manufacturers do not include such information in their catalogs, they should be contacted to obtain it.

Check valve sizing may result in a valve of a smaller size than the pipeline in which it will be installed. This is not necessarily undesirable. A smaller, fully open check valve usually will have no greater resistance to flow than a partially open one that is the same size as the pipeline, with the advantage that it will not wear as quickly. In addition, the lower cost of the smaller valve will offset the cost of the pipe reducers required to fit it into the line.

5

Butterfly Valves

In the butterfly valve the flow control element, which is a circular disc, is rotated about an axis that is perpendicular to its own axis and to the fluid path. It is always in the fluid path, and when the valve is open it splits the flow into two separate paths around it. Because the butterfly valve is essentially symmetrical, either end can be the inlet, and thus flow can be in either direction. The form of control for which butterfly valves are best suited is regulated flow volume (called *throttling*); however, they can also be used as stop valves (for starting and stopping flow).

The butterfly valve is one of a group of valve types (the ball valve and the plug valve are the others) commonly called *quarter-turn valves*. This is because their flow control element moves from the fully open position to the fully closed position with just 90° shaft rotation. Most butterfly valves are one of two designs, the lined and the high performance.

LINED BUTTERFLY VALVE

An example of the lined butterfly valve is shown in Figure 5-1. It consists of three major components: the body, the disc, and a non-metallic body liner. The body is a relatively simple ring of metal. It is usually lug-style, as shown

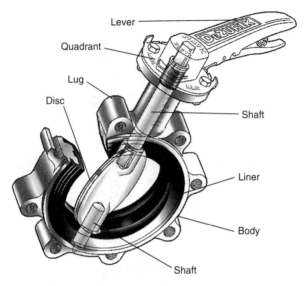

Figure 5-1. Lined butterfly valve

in Figure 5-1, or wafer-style, as shown in Figure 5-2. These "endless" body styles have been discussed in Chapter 1. Bodies are also made with flanged ends.

The lined butterfly valve has a conventional, or concentric, disc. It pivots on a single shaft, which passes through its center, or on two partial shafts, also centered, as shown in Figure 5-2. The shaft is pinned to the disc. Its ends pass through the liner and are supported by the body, typically in metallic bushings. One shaft end passes completely through the body where a lever, used to operate the valve, is attached. The lever is oriented on the shaft parallel to the disc so that the position of the lever also indicates the position of the disc. The disc itself has a tendency to turn on its own in response to the force of the flowing fluid. To provide positive disc positioning, a quadrant is fastened to a small flange on the body where the shaft passes out of the body. The quadrant has a series of notches (typically at 10° increments) around its perimeter. The lever used to rotate the disc is provided with a detent mechanism that can engage one of the notches on the quadrant, causing the lever, shaft, and disc to be held in position. Where abrasion or corrosion resistance is needed, the disc of the lined butterfly valve is readily coated with an appropriate non-metallic material.

The resilient non-metallic liner is bonded to the body or held in grooves in the bore of the body. The liner performs several important functions.

1. It provides the body seating surface. The liner is crowned at the center; therefore, when the valve is closed the edge of the disc presses tightly against the liner, compressing it and forming a seal around almost the entire perimeter of the disc. The shaft ends prevent complete circular seating of the disc.
2. It fits tightly against the disc hubs and the shaft where the ends pass through the liner, forming a seal for them. In addition to the designed-in gripping action of the liner around the shaft ends, O-ring seals are frequently provided at the liner–body interface to protect the body and prevent external leakage.
3. Its faces extend slightly beyond the ends of the body and serve as a gasket for the connecting pipe flanges.
4. It protects the valve body from the fluid and permits the selection of body materials based only on strength and price, without concern for abrasion, erosion, and corrosion resistance.

The lined butterfly valve has several drawbacks. First, it is difficult to maintain a seal between the disc and the liner in the area adjacent to the ends of the shaft. This area is subject to wear, because the disc is always in contact with it and rubbing occurs whenever the disc is moved. Second, because the compression seal between the edge of the disc and the liner is difficult to maintain at high pressures, the lined valve is only suitable for low fluid pressures. Finally, the allowable service temperatures of non-metallic body liner materials limit this design to low fluid temperatures.

Lined butterfly valves are manufactured in cast iron and ductile iron, with liners made of non-metallic materials such as Buna-N, EPT and TFE. (Non-metallic materials are discussed in Chapter 9.) They are available in sizes from 2 inches to 24 inches and are capable of handling fluid pressures of up to 175 psi and temperatures from –20°F to 400°F, depending on liner material.

HIGH-PERFORMANCE BUTTERFLY VALVE

The high-performance butterfly valve design shown in Figure 5-2 eliminates two drawbacks of the lined butterfly valve, seat wear and seat leakage, and when furnished with a metal seat ring, also eliminates the third. Its major components are the body, disc, and seat ring. The body is similar to that of the lined valve, but its bore is machined to receive the seat ring. It is also made in wafer and lug style, and with flanged ends.

Unlike the symmetrical conventional disc of the lined valve, the high-performance valve uses an off-set (or *eccentric*) disc. It is shown schematically in Figure 5-3. In this design the center of rotation of the disc is off-set from the centerline of the disc in two directions. The first off-set is axial. The shaft passes through the disc behind the line of

Butterfly Valves

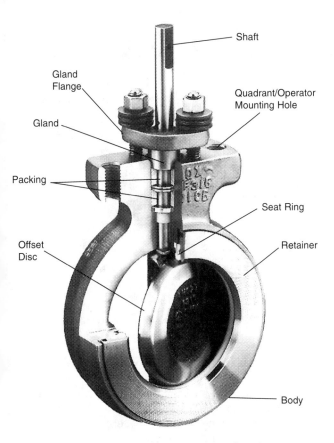

Figure 5-2. High-performance butterfly valve

60 **Butterfly Valves**

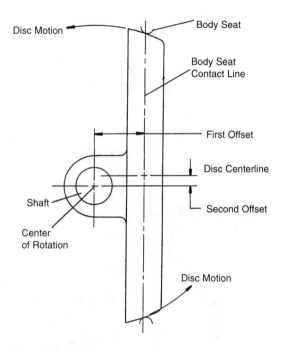

Figure 5-3. High-performance valve disc offsets

contact and between the disc seating surface and the seat ring. This leaves both seating surfaces uninterrupted and achieves a seal for the full 360° around the disc. The second off-set is lateral. The center of the shaft is to the side of the centerline of the disc. This off-set, coupled with the tapered disc seating surface, causes the disc to pull (or *cam*) away from the seat ring just as the valve opens and to cam into the seat ring just as the valve closes. Wear of the seat ring is minimized, and opening and closing forces

Butterfly Valves

are reduced. This design also makes it possible to replace the body seating surface without disassembling and removing the disc and shaft.

The disc is positioned in the body by shaft spacers between the disc and the body. These spacers are usually backed up by seals to prevent the fluid from coming into contact with the shaft bearings. To prevent external leakage at the pressures at which these valves are used, the extended shaft of the high-performance butterfly valve is also provided with a stuffing box, a packing gland, and a gland flange. These parts are shown in Figure 5-2 and have been discussed in Chapter 2. High-performance butterfly valves also use the lever-and-quadrant arrangement for disc operation and positioning.

In the high-performance butterfly valve, the seat ring can be made either of non-metallic material (called a *soft seat*) or of metal. Both are designed to flex when contacted by the disc when closing, forming a tight seal. However, the typical metal seat ring does not achieve a completely leak-tight seal. With the soft seat, when the valve is installed so that the inlet is at the seat-ring end of the valve (in the normal direction), fluid pressure on the seat ring forces it hard against the tapered disc seating surface, enhancing the seal. As fluid pressure increases, the seal becomes tighter. Thus the high-performance valve can be used at higher pressures than can the lined valve. When flow is from the shaft side of the disc, the seat ring retainer prevents the ring from being pushed away from the disc, and thus the seal is maintained. Also, because it has a smaller inside diameter than does the seat ring, the retainer also protects the seat ring from abrasion and erosion when the valve is installed conventionally.

High-performance butterfly valves are manufactured in ductile iron and in carbon, alloy, and stainless steels in sizes from 2 inches to 72 inches. These valves can handle fluid pressures of up to 1,500 psi with non-metallic seats and pressures well over 2,500 psi with metal seats. Fluid temperatures can be accommodated from 50°F to 500°F with non-metallic seats, which usually are made of TFE, and over 1,000°F with metal seats, which usually are made of stainless steel.

APPLICATIONS

The butterfly valve can be used as a stop valve; however, it is primarily a throttling valve. The lined butterfly valve, with its conventional disc, exhibits a flow characteristic known as *equal percentage:* When the valve opening and fluid flow through the valve are small, a change in disc position produces a small change in flow rate; conversely, when the valve opening and fluid flow through the valve are large, the same increment of change in disc position produces a large change in flow rate. The result is that flow control at large valve openings is less sensitive, and the lined butterfly valve is generally limited to 60° disc rotation.

The high-performance butterfly valve exhibits a flow characteristic that lies between the equal percentage and the linear flow characteristics. The change in fluid flow is not directly proportional to the change in disc position throughout the disc travel, as it would be with the linear flow characteristic. However, the change in flow for equal increments of disc movement becomes progressively less as the valve opens. This characteristic causes the high-performance valve to remain sensitive to required flow changes throughout its full 90° disc travel, and makes it preferable to the lined valve for many throttling applications. (Flow characteristics are discussed in greater detail in Chapter 11.)

Butterfly Valves

Whether used as a throttling valve or a stop valve, the butterfly valve can be used with liquids, gases, and vapors. It also has some definite advantages over the other valve types:

1. It is available in larger sizes than other valve types—up to 72 inches. Standard gate valves, for example, are available only up to about 30 inches and globe valves only up to 16 inches, unless they are specially fabricated. The availability of very large sizes makes the butterfly valve almost the only choice for very large fluid flow rates at low and moderate pressures.
2. Its wafer- and lug-style bodies make the butterfly valve shorter end-to-end, lighter, and frequently less costly than the same size valve of another type.
3. Its quarter-turn operation makes it fast-acting. It is especially useful as a stop valve when frequent operation is required.
4. The lined valve can be used with liquid slurries and gases carrying suspended solids and solid particles because it has no pockets in which solids can collect. When the disc is coated with a suitably resilient material, the lined butterfly valve can be used with abrasive materials.

As do other valve types, the butterfly valve has limitations. It is not made in small sizes (under 2 inches). The lined valve is very limited as to the fluid temperatures and pressures it can handle. With soft seats, the high-performance butterfly valve has fluid-temperature limitations. With metal seats, the temperature and pressure capabilities of the high-performance valve compare with those of other valve types; however, the metal-seated valve does not achieve leak-tight closing, making it undesirable for use as a stop valve. In addition, when used as a stop valve the butterfly valve has a higher resistance to flow when open than do other stop valves, because the disc is always in the fluid path. Finally, the force required to operate a large butterfly valve is great, making it necessary to use an operator or actuator, thereby increasing total valve weight and cost. (Operators and actuators are discussed at length in Chapter 12.)

6

Ball Valves

The flow control element of a ball valve is a ball with a round hole through it. The ball rotates about an axis that is perpendicular to the hole and to the fluid path. The ball is always in the fluid path. However, when the valve is open the hole aligns with the fluid path, and the fluid passes straight through it. Because the ball valve is symmetrical, either end can be the fluid inlet, and thus flow can be from either direction. The form of control for which ball valves are best suited is starting and stopping flow; however, they also can be used for changing flow direction.

BALL VALVE DESIGN

Figure 6-1 shows a typical ball valve. It is a "split body, flanged end, floating ball, full port, soft-seated, ball valve." These terms describe the particular design features of the valve with respect to body construction, end connections, ball design, and seat design, in that order. End connections have been covered in Chapter 1 and will not be discussed here. Body construction, ball design, and seat design are discussed below.

1. Body Construction. Unlike the other types of valves that may have substantially different designs but that are assembled in the same way (e.g., the knife gate

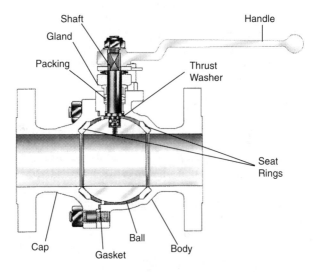

Figure 6-1. Split body ball valve

Ball Valves

valve versus the standard gate valve), the ball valve has a single design configuration with different designs based on how the ball is inserted (or *loaded*) into the body. The different designs are the split body, end entry, top entry, and 3-piece.

- The split body design shown in Figure 6-1 consists of two body parts, the body and the cap, that have mating flanges held together with studs or bolts and nuts. The split between the parts is off-center so that the stuffing box area is left intact, simplifying sealing of the shaft and of the body cap. A seat ring, the shaft, and the ball are inserted into the body, in that order. Then the cap, which has a seat ring in place, is bolted on. A gasket is placed between the body and cap flanges to prevent external leakage. On small valves it is common to have the two parts screwed together, with the cap being screwed into the body.

  Split body construction is available in valve sizes from ½ inch to 36 inches. Flanged ends are available on all sizes of split body valves and are the standard for large valves. Threaded ends are standard on small valves ½ inch to 2 inches.

  The body joint in the split body design is a potential source of leakage. On small, disposable, screwed assembly steel valves, the joint is sometimes seal-welded.

- The end entry design is shown in Figure 6-2. The assembly of this design is similar to that of the split body design, except that the body is a single piece. The ball is installed from the appropriate end and is retained by an insert screwed into the body. End entry construction is used on ½-inch to 6-inch flanged end valves. It is also stan-

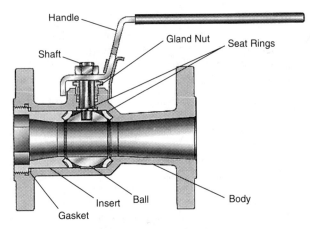

Figure 6-2. End entry ball valve

dard on small, inexpensive threaded end valves, sizes ½ inch to 2 inches, because the same threads used to install the valve in the pipeline can be used to install the insert into the valve. Because the body is made of one piece in this design, one source of external leakage is eliminated.

- Figure 6-3 shows the top entry design, which also has a one piece body, with the ball inserted from the top. In the valve in Figure 6-3, to insert the ball into the body, the ball must be made with portions of its top and bottom removed. Assembly of the valve begins with installation of the two seat rings. Then the ball, with its hole directed vertically, is dropped into the body and rotated 90° in the vertical plane so that the hole aligns with the fluid path. Finally, the bonnet, with the shaft in place, is bolted to the body. A gasket is used to prevent leakage at the body–bonnet joint. There are two advantages to this design. First, the top entry ball valve can be serviced by removing the bonnet, without removing the valve from the pipeline. Second, top entry valves with welding ends can be welded to the connecting pipe without temperature-sensitive, non-metallic seat rings of the top-entry valve being in place.

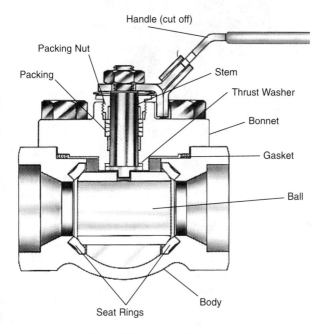

Figure 6-3. Top entry ball valve

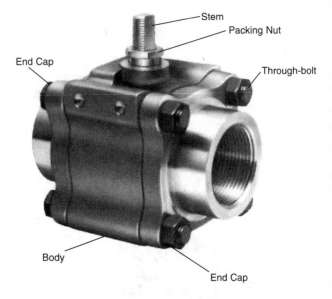

Figure 6-4. 3-piece ball valve

Top entry construction has been used on valve sizes up to 24 inches but is more common on smaller valves, up to 4 inches. It is available with threaded, socket-weld, butt-weld, and flanged ends.

- The 3-piece design is shown in Figure 6-4. In this design the shaft, ball, and seat rings are inserted into the body, in that order. Then two end caps are fastened to the body with bolts that pass through holes in all three pieces, as shown. On larger valves the caps are held on by studs in the body and nuts. Gaskets situated between the body and end caps prevent leakage.

In line servicing and damage free welding are the advantages of this design. Its disadvantages are the potential leak paths at the body–cap joints. 3-piece construction is used on ball valve sizes that range from ½ inch to 36 inches and is available with threaded, socket-weld, flanged, and butt-weld end connections.

In all these body constructions the shaft is separate from the ball and is inserted into the body or bonnet from the inside, before the ball is inserted. The body or bonnet has a stuffing box to hold packing or seals that are held in place by a gland and gland flange or a gland nut. The shaft has an integral collar that retains it in the body or bonnet and keeps fluid pressure from causing shaft *blowout*. A thrust bushing between the shaft collar and the body or bonnet reduces friction and wear when the shaft is turned. The top of the shaft is fitted with a handle to turn the ball. The handle lines up with the axis of the ball hole so that the position of the handle shows the position of the ball. Stops on the top of the valve body or bonnet limit movement of the handle to 90°.

2. Ball Design. There are two aspects to ball design, port size and the manner of ball support. Port size is the diameter of the hole in the ball. Three different port sizes are commonly available: full, regular (or standard), and reduced. The diameter of the full port ball design is equal to the size of the valve and is approximately the same as the inside diameter of standard-weight pipe of the same size as the valve. Full port size offers the least resistance to fluid flow; however, it is not feasible for all body constructions and produces large end-to-

end valve dimensions. The diameter of the hole of the regular port ball is smaller than that of the full port ball. The amount of reduction varies with valve size but is approximately 75% to 90% of the full port diameter. Although this increases the resistance to flow, it enables the ball and the valve to be smaller and lighter than the full port ball and valve. Regular port ball valves are

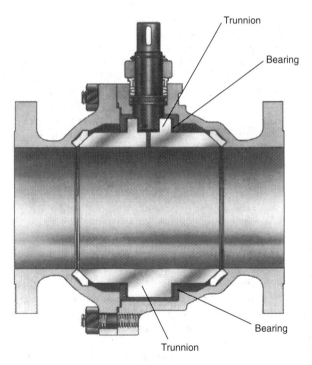

Figure 6-5. Trunnion ball valve

generally shorter end-to-end than are full port valves. They are also made to be interchangeable with the same size gate valves. The weight difference between full port and regular port valves is substantial in large valves. The reduction in hole diameter of reduced port balls is approximately 60% of the full port balls. The reduced port is found on small, threaded end, end-loaded valves in which the body construction requires a small ball diameter.

There are primarily two ways to support the ball in the valve body: the floating-ball design, shown in Figures 6-1 to 6-4, and the trunnion-mounted ball design, shown in Figure 6-5. In the floating-ball design the ball is supported only by the two seat rings. A slot in the top of the ball that is perpendicular to the hole accepts the shaft, which is machined to fit into the slot. This arrangement enables the shaft to turn the ball and also allows the ball to move (or *float*) in the direction of the slot. Consequently, when the valve is closed upstream, fluid pressure pushes the ball against the downstream seat ring, enhancing the seal. Depending on the seat design, movement of the ball downstream may relieve the seal at the upstream seat ring. In the trunnion-mounted ball design the ball is made with integral short-shaft extensions (or *trunnions*) at the top and bottom. These trunnions fit into bearings that, in turn, are held in the body or in transition pieces that are inserted into the body. The shaft fits into a slot in the top trunnion, or a splined joint is used. In this design the ball is held firmly in place and is not moved by fluid pressure. The tightness of seal is solely dependent on the seat design.

The floating ball is used with all the different ball valve body designs; whereas the trunnion-mounted ball design is used mainly with the split body design. The torque required to turn a trunnion-mounted ball is significantly less than that needed for a floating ball; therefore, the trunnion-mounted ball design is standard for large split body valves.

3. Seat Design. The different seat designs are based on the material used to make the seat rings. There are two classifications: "soft" seats made of resilient, nonmetallic materials such as TFE, PFA, and PEEK; and metal seats made of stainless steel or base metals coated with Stellite® or tungsten carbide. (These materials are discussed in Chapter 9.) Properly designed soft seat rings are very effective and produce leak-tight seals that are referred to as *bubble tight;* however, seat rings are limited by the service temperatures of the seat materials. Metal seats can withstand higher temperatures but are not usually as tight as soft seats.

There are two designs of soft seat, the "jam" seat and the flexible seat. In both designs the resilient seat rings are placed into circular recesses in the body and seal against both the ball and the body. With the jam seat the rings are mechanically compressed by the ball during assembly. This can produce a tight seal initially; however, the design is not good for services with wide pressure or temperature fluctuations. The resilient materials used for the seat rings have a relatively small elastic range; therefore, it is not difficult for them to be compressed into their plastic range, experience *cold flow* (that is, permanent deformation), and cause the loss of the initial seal pressure. Overcompression can occur at assembly if part tolerances, assembly clearances, or both are not closely controlled. Overcompression can occur with high fluid pressures, and because the rate of thermal expansion is much greater

Ball Valves

for the resilient material than for the metal body and ball, it can be induced by high temperatures. The jam seat is found only on small, inexpensive ball valves.

The limitations of the jam seat are overcome by the flexible seat. In this design the seat ring either is made with a lip that flexes when contacted by the ball during assembly, or the ring is arch-shaped in cross-section and flexes at assembly. The lip-type seat ring is shown in Figure 6-6, and the arched seat ring is shown in Figures 6-1 to 6-5. The stresses produced by flexing are substantially lower than those produced by compression for the same seat ring contact surface displacement. Therefore, cold flow is eliminated, and the ring maintains its sealing pressure even with temperature and pressure fluctuations. The flexible seat is used with all sizes, body designs, and ball designs of ball valves.

Metal seat rings are not inherently resilient as are soft seat rings. To compensate for this, springs are commonly used to maintain contact pressure between the seat rings and the ball. This is especially true with trunnion-mounted balls in which both seat rings must be backed up with springs. In the floating ball design, only one seat ring needs to be backed up with a spring. If flow is always in the same direction and seating is acceptable only at the downstream seat, then springs may be omitted altogether in the floating ball design.

Unlike the soft seat, metal seats are not effective in sealing between the seat ring and the body. Therefore, to prevent leakage a backup seal at the ring–body interface is required. If temperature is not a consideration, the backup seal can be made of the same material as is a soft seat ring. For higher-temperature services, graphite or flexible metal rings that also act as springs are used. Metal seat rings enable ball valves to be used

Ball Valves

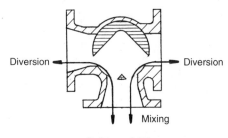

Ball turned 45°

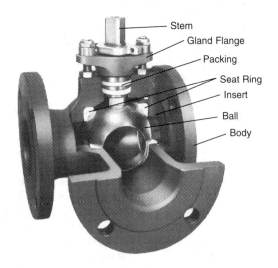

Figure 6-6. 3-way ball valve

at temperatures higher than those at which resilient seats can be used. In addition, metal seat rings are particularly suitable for applications in which abrasive solids are suspended in the fluid. They usually are available only on large split body, top-entry, and three-piece valves.

BALL VALVE APPLICATIONS

Ball valves find application mainly as stop valves. As a stop valve it has the following attributes:

1. It has a smooth, uninterrupted flow path that offers little resistance to flow when it is open. This enables the ball valve to be used with slurries, solids suspended in gases or air, and dry powders without concern for build-up of solids.
2. It is manufactured in a wide range of sizes, pressure classes, and materials. Sizes range from ½ inch to 36 inches. Ball valves can handle fluid pressures from vacuum to over 2,500 psi, and they are available in most types of valve materials: bronze, steel, and corrosion-resistant alloys. They are not usually made of cast iron.
3. It can be used with all fluids, liquids, gases, and vapors (including steam), whether clean or dirty. Fluid temperatures are limited only by the materials of construction.
4. Its quarter-turn operation and low required operating force make it preferable for applications in which frequent operation is required, and make it easy to actuate.

The ball valve does have some drawbacks. The full port valve is large from end-to-end and takes up more pipeline space than does a gate valve of the same size, especially in very large sizes. With soft seats, ball valves have fluid-temperature limitations. With metal seats, there are no temperature limitations; however, leak-free seating cannot be assured, and the cost of the valve is higher.

Ball valves can be used in throttling service. They have an equal percentage flow characteristic; therefore, the fluid flow rate is not proportional to the ball position. (See Chapter 11 for a discussion of flow characteristics.) In addition, the fluid must be non-abrasive, because the downstream seat ring is directly in the flow path and exposed to the flowing fluid when the valve is not completely open.

THREE-WAY BALL VALVE

The ball valve is one of the two valve types that can be used to change flow direction. (The plug valve is the other.) Figure 6-6 shows a ball valve designed to be used for changing flow direction. It is called a three-way, two-port valve, identifying the number of body openings (three) and the number of ball openings (two). It has a single-piece, end entry body with an added side connection. The ball is not bored all the way through, but has two intersecting, partial depth holes 90° apart. See Figure 6-6. In all other respects the valve is identical to a standard two-way, end entry ball valve. In operation, the ball can be turned so that either of the body end openings can be connected with the side opening, with flow either into or out of the side opening. The ball can also be turned only 45°, connecting both the end openings to the side opening for *diversion* (flow into the side opening and out both ends) or *mixing* (flow in both ends and out the side). This is shown at the top of Figure 6-6. With the ball turned 45° the seat rings are exposed to the fluid; therefore, abrasive fluids should be avoided.

7

Plug Valves

The flow control element of the plug valve, the *plug*, is a cylinder or a truncated cone with a hole through it. The plug rotates about an axis that is perpendicular to the hole and to the fluid path. The plug is always in the fluid path. However, when the valve is open the hole aligns with the fluid path, and the fluid passes straight through it. Because the plug valve is symmetrical, either end can be the fluid inlet, and thus flow can be from either direction. If the plug is cylindrical, the hole may be rectangular or round. If the plug is a truncated cone, the hole is usually a trapezoid.

The forms of control for which the plug valve is suited are starting and stopping flow and changing flow direction.

PLUG COCK

A basic plug valve, commonly called a *plug cock,* is shown in Figure 7-1. Construction of this valve is quite simple. It has only three major parts: the body, plug, and bottom cap. There are no separate seat rings. The outer surface of the tapered plug bears directly against the tapered bore in the body, which is the body seat. Leak-tightness between the plug and body is maintained by an adjusting screw or by a spring bearing against the large end of the plug, as shown in Figure 7-1. There is no separate shaft to turn the plug. The small end of the plug extends outside the body and has flats machined on it to

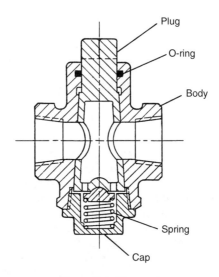

Figure 7-1. Plug cock

accept a removable wrench. (Plug valves usually are furnished without a lever handle.) Sealing against external leakage is accomplished by the plug retaining cap and an O-ring at the small end of the plug.

The plug cock design has deficiencies. Leak-tightness depends primarily on the quality of the machined sealing surfaces on the plug and body. It becomes more difficult to machine good, leak-tight surfaces as the size of the valve increases, and even relatively well-fitted sealing surfaces have difficulty sealing at high pressures. Forcing the plug into the body to improve tightness only serves to aggravate another weakness, that is, the plug becomes harder to turn. The large bearing surface between the plug and body causes the amount of force needed to turn the plug to be great, and the metal-to-metal contact can result in seizing and galling. This is especially true if the valve has not been operated for a long time. Finally, as valve size increases, the required operating force increases to an impractical amount.

These deficiencies restrict plug cocks to small sizes and low-pressure applications. Plug cocks are made of bronze and cast iron in sizes from ¼ inch to 2 inches with threaded ends. Plug cocks can handle pressures only of up to 250 psi.

To correct the deficiencies of the plug cock and to improve the range of plug valve sizes and pressure capabilities, other designs have been developed. Two of these are the lubricated plug valve and the sleeved plug valve.

LUBRICATED PLUG VALVE

A typical lubricated plug valve is shown in Figure 7-2. This design is very similar to the plug cock but with an important difference—the plug. The plug is drilled from the top of its integral shaft along its centerline and then radially from the cylindrical seating surface, intersecting the centerline hole. The seating surface is grooved circumferentially and longitudinally. A sealant injection fitting screwed into the hole at the top of the plug is used with a sealant gun to inject sealant down the center of the plug, out the radial holes, and into the surface grooves from which it oozes out, filling the narrow clearance between the plug and body seating surfaces. A small check valve just below the injection fitting prevents the sealant from leaking back out. A spring at the bottom of the plug presses the top of the plug against a gasket held in the top of the body, preventing external leakage.

The sealant performs three functions: it forms a renew-

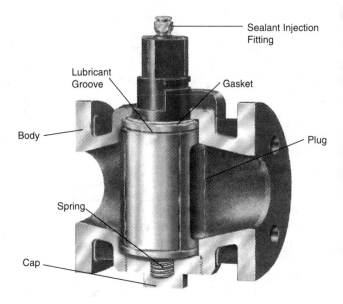

Figure 7-2. Lubricated plug valve

able seal between the plug and body, preventing internal leakage; it protects the seating surfaces against corrosion; and it acts as a lubricant, reducing the force required to operate the valve. A slightly different design uses sealant to further reduce the operating force. It has a conical plug that is inserted into the body from the top and held in place with a cover. The lubricant grooves in this plug extend all the way to the bottom of the plug so that sealant fills the cavity between the bottom of the plug and the body. If sealant is injected immediately before turning the plug, the additional pressure of the sealant on the bottom of the plug raises the plug slightly, increasing the clearance between the plug and the body and thereby reducing the turning force.

Manufacturers of lubricated plug valves offer a selection of sealants to use with different media: air, acids, alkalis, alcohols, water, steam, petroleum oils and fuels, and so on. Some sealants are more versatile than others; however, no single one is suitable for all fluids. Sealants must be matched to the fluid to ensure that it does not dissolve the sealant and "wash" it out. This causes the valve to leak and the plug to seize and gall when turned. Service temperatures for sealants range from about −20°F to 450°F, precluding use of the lubricated plug valve at temperatures above 450°F. In addition, in some services the sealant could contaminate the process fluid.

Lubricated plug valves are manufactured in bronze, cast iron, and carbon steel in sizes from ½ inch to 36 inches and with the capability of containing fluids at pressures well over 2,500 psi.

SLEEVED PLUG VALVE

A sleeved plug valve is shown in Figure 7-3. Rather than using sealant to reduce friction and achieve sealing, a non-metallic sleeve is locked in recesses in the body. The polished, slightly tapered plug acts as a wedge, pressing against the sleeve, and a tight seal is obtained. The plug is forced against the sleeve by a thrust collar acting through a non-metallic gasket at the top of the plug. The force of the thrust collar can be increased by tightening several adjusting bolts that pass through the valve cover. The gasket also seats against the integral plug shaft to prevent external leakage. Sleeve materials are selected from those with low coefficients of friction, reducing the required turning force; however, the forces are higher than those of the

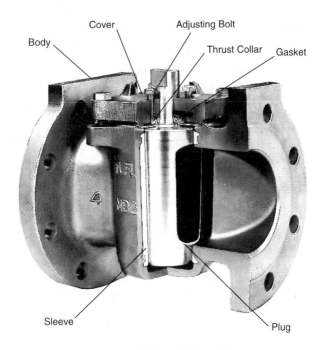

Figure 7-3. Sleeved plug valve

same size lubricated valve. Also, the non-metallic sleeve has temperature limits that are about the same as the sealant for the lubricated plug valve.

Sleeved plug valves are manufactured in a wide range of valve materials: cast iron, carbon steel, stainless steel, and the more exotic nickel-based alloys. Sizes range from ½ inch to 18 inches. Sleeved plug valves can contain fluid pressures up to 1,500 psi. Sleeves are made primarily of

TFE or TFE derivatives. (See Chapter 9 for a discussion of these different materials.)

PLUG VALVE PORTS AND PATTERNS

Plug valves are made with different size port openings, which affect plug and body size. Plug valves with openings that have an area equal to that of the connecting pipe are called 100% ports. These 100% ports can be round, rectangular, or trapezoidal, depending on the shape of the plug. The 100% round port results in the largest plug valve. Its use is preferred where resistance to flow must be kept to a minimum and where the valve is used in a pipeline that is internally scraped (or *pigged*). The 100% round port is only available with a cylindrical plug. The 100% rectangular port is shown in Figure 7-2. To match the opening height, the body flow passages are tapered. The 100% trapezoidal port valve is similar. The 100% port plugs are supplied in regular pattern bodies, which have the largest end-to-end dimensions of the different body designs.

The most common plug valves are those with *regular* ports, which are port openings that vary from 40% to 100% of the connecting pipe area, depending on valve size. Most openings range from 60% to 70% of the connecting pipe area. Regular port plugs are supplied in both regular and short pattern bodies. Short pattern body end-to-end dimensions match the end-to-end dimensions of the same size gate valve; regular pattern body valves are longer.

The smallest port opening plugs are found in venturi pattern bodies in which the flow paths of the body taper down from the valve ends to the plug. Venturi plug port openings are 40% to 50% of the connecting pipe area. Venturi pattern bodies have the same end-to-end length as the regular pattern. They are used on large valves (6 inches and larger) to reduce size, weight, and operating torque when higher resistance to flow is acceptable.

APPLICATIONS

Plug valves are used as stop valves and to change flow direction. Changing flow direction is accomplished with multi-port plug valves. These will be discussed separately in the next section of this chapter. As a stop valve, the plug valve has the following attributes:

1. It is a simple design, with few parts, and usually may be serviced without removal from the pipeline.
2. It has a smooth, uninterrupted flow path that offers little resistance to flow when open.
3. It is manufactured in a wide range of sizes, pressure classes, and materials. Sizes range from ½ inch to 36 inches. Plug valves can handle fluid pressures from vacuum to over 2,500 psi, and they are made of all available types of valve materials: bronze, cast iron, steel, and corrosion-resistant alloys.
4. It can be used with all fluids, liquids, gases, and vapors (including steam), whether clean or dirty. Owing to its uninterrupted flow path, the plug valve is well suited for handling slurries of coal and mineral ores, muds, sewage, and media that cause scale or precipitates to form. Plug valves are also frequently used in gas transmission pipelines and in oil fields in the production of crude oil.
5. Its quarter-turn operation and low required operating force make it preferable for applications in which frequent operation is required, and make it easy to actuate. The lubricated plug valve typically has a lower required operating force than any other type of valve of the same size.

6. The seat of a lubricated plug valve can be renewed by the injection of sealant.

As do all valve types, plug valves have drawbacks. The regular pattern body valve is large from end-to-end and takes up more pipeline space than does the same size gate valve, especially in very large sizes. Both the lubricated and the sleeved valve design have temperature limitations owing to the allowable service temperatures of sealants and non-metallic sleeve materials. In addition, unlike other valve types, the lubricated plug valve requires periodic maintenance or an additional step before operating—the injection of sealant. If the sealant is not renewed periodically, fluids may dissolve it and wash it from the plug face, causing seizing and galling between the plug and the body when the plug is turned.

MULTI-PORT PLUG VALVES

The plug valve is one of two valve types that can be used to change fluid flow direction. (The ball valve is the other.)

All three designs discussed in this chapter can be modified to be multi-port valves. To accomplish this modification, both the body and the plug are changed from those in the standard plug valve. One or two end connections are added to the sides of the body, and the plug has either an additional flow passage that is perpendicular to and intersects the normal passage, or two nonintersecting flow passages that do not go all the way through the plug but turn 90° and exit it. With the sleeved plug valve, it is also possible to add a connection at the body to work in conjunction with a flow passage in the plug that starts at the bottom and then turns 90° to exit at the seating surface of the plug. Multi-port valves are designated by the number of

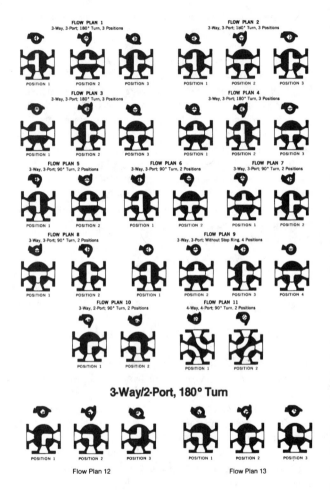

Figure 7-4. Multi-port plug valve designs

body openings (called *ways*) and the number of plug openings (called *ports*). In multi-port valves, plug rotation is not limited to 90° but may be a full 360°.

There are many combinations of ways, ports, and plug rotation that produce different schemes of change in flow direction. One manufacturer's standard offerings are shown in Figure 7-4, which also shows the body and plug modifications described previously.

Multi-port plug valves are useful in many installations and allow for simpler and more compact piping. This results in a savings in valve, pipe, and pipe fitting costs, and the cost of labor for installation. These valves also add convenience and safety to operations. Operating one valve can simultaneously disconnect one process and connect another. The risks of overlooking the opening or closing of one valve of several and the serious possible consequences are avoided.

Multi-port plug valves are used in such applications as blending fluids, diverting fluid from one destination to another; switching between parallel pieces of equipment, such as pumps and filters; and sequencing between different processes. These applications are shown schematically in Figure 7-5. The valve shown in the blending application is a three-way, three-port valve. The plug is in the center position, and fluid-supply tanks, Tank 1 and Tank 2, are connected to the process. Turning the plug 90° clockwise or counterclockwise connects only Tank 1 or Tank 2 to the process. The valve shown in the diverting application is a three-way, two-port valve. The fluid-supply tank is connected only to Process 1. If the plug is turned 90° clockwise, the tank is connected only to Process 2. In the switching application, two three-way, two-port valves are used. The fluid passes through Unit 1, which may be a pump, a filter, and so on. Turning the

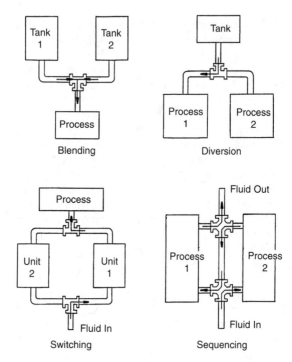

Figure 7-5. Multi-port plug valve applications

inlet valve 90° clockwise and the outlet valve 90° counterclockwise causes the fluid to flow through Unit 2 and disconnects Unit 1 for cleaning or repair. In the sequencing application, two four-way, four-port valves are used. The fluid passes first through Process 2 and then through Process 1. By turning both valves clockwise (or counterclockwise), the fluid passes through Process 1, then Process 2. In the switching and sequencing applications,

Plug Valves

the valves may be coupled so that both plugs can be moved simultaneously.

When applying multi-port valves, it is recommended that they be installed so that the plug is between the pressure-inlet body opening and any body opening that is closed off. This is called a *positive position.* With the plug in a positive position, fluid pressure tends to push the plug against the closed port and assists in sealing the port. If the valve is installed so that fluid pressure is in a closed body opening and against the surface of the plug (called a *negative position*), the fluid pressure tends to push the plug away from the closed opening. When the valve is used with the plug in the negative position, only moderate pressures are recommended. Furthermore, a tight shut-off is not ensured. In general, multi-port plug valves are intended for directional control only and are not effective in sealing high differential pressures between ports.

8

Diaphragm Valves

The flow control element of a diaphragm valve is a flexible membrane, called a *diaphragm,* that is deformed to close the fluid path. The diaphragm lines the side of the fluid path and is pushed across it until the diaphragm seats against the other side of the fluid path, stopping flow. Because the diaphragm valve is symmetrical, either end can be the inlet, and thus flow can be from either direction through the valve. The form of control for which diaphragm valves are suited is as a stop valve, that is, for starting or stopping flow, or regulating flow volume (called *throttling*).

DIAPHRAGM VALVE DESIGN

Almost all diaphragm valves are one of two very similar designs, the straight-through and the weir design. The weir design is also known as a Saunders valve, named after the original patent holder.

An example of a straight-through diaphragm valve is shown in Figure 8-1. The design is called straight-through because when the valve is open the fluid path through it has the same size and shape as the connecting pipe. It consists of just three major components: the body, the diaphragm, and the bonnet. The valve shown in Figure 8-1 is flanged; however, diaphragm valves are available with all the common valve ends: threaded, socket-weld, flanged, and butt-weld. The body does not have a separate

Diaphragm Valves

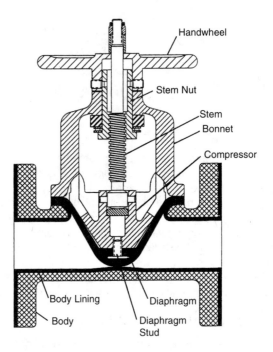

Figure 8-1. Straight-through diaphragm valve

seat but is appropriately contoured to provide a compatible seating surface for the diaphragm. The flanged body can be fully lined with non-metallic materials, as illustrated in Figure 8-1, to provide it with additional abrasion or corrosion resistance.

Diaphragms are made of flexible non-metallic materials. During manufacture they are formed to the shape they will have when the valve is closed. They are not stretched into shape when the valve is in the closed position. When the valve is opened, the diaphragm flexes as it is pulled back into the bonnet. Fastened to the diaphragm by a molded-in diaphragm stud is an intermediate member called a *compressor*, which is shaped to back up the diaphragm when the valve is in the closed position and prevent its distortion from the fluid force. In addition to being the flow control element, the diaphragm contains the fluid pressure and seals the bonnet and all its contents from the fluid passing through the body. It also acts as a gasket between the body and the bonnet, preventing external leakage. The diaphragm may be fabric-reinforced or laminated of different non-metallic materials to give it the desired overall properties of flexibility, strength, and corrosion resistance.

The bonnet is protected from the fluid and its pressure; therefore, its primary function is to house the mechanism that moves the diaphragm. The diaphragm is moved by a non-rotating stem, which is pinned to the compressor. The threaded stem mates with a stem nut at the top of the bonnet. The stem nut is free to turn but is held in place by a flange at its bottom and the valve handwheel at the top. Turning the handwheel turns the stem nut and causes the stem to move into or out of the valve. The thread on the stem is left-handed, so that turning the handwheel clockwise moves the stem downward, closing the valve. Because the diaphragm seals the entire bonnet, a stem seal is not required. However, for valves in systems that handle toxic and dangerous fluids—in which diaphragm failure would be disastrous—a stem design with packing, as shown in Figure 2-7, can be provided. Also, because the bonnet is not exposed to the fluid, it need not be made of the same material as the valve body. It is not uncommon for valves with expensive corrosion-resistant-alloy bodies to have bonnets made of inexpensive cast iron.

A drawback of the straight-through valve design is the limited number of materials that are able to tolerate the amount of flexing required of the diaphragm. Only natural rubber and some of the elastomers (synthetic rubbers) such as Hypalon® and Neoprene are suitable. (See Chapter 9 for a discussion of non-metallic materials.) As a general rule, the more chemically inert a non-metallic material is, the less flexible it is; and the more flexible it is, the less chemically resistant it is. Consequently, the straight-through design has limitations as to the fluids it can handle and their temperatures.

The design of the weir valve shown in Figure 8-2 overcomes this limitation. The transverse weir (or *dam*) formed into the valve body or body lining reduces the required diaphragm travel from the closed valve position to the open valve position so that diaphragm flexing is greatly reduced. As a result, a larger selection of materials can be used. All of the common elastomers and even some plastics, such as highly chemically resistant TFE, can be used for diaphragms. In all other respects the weir design is essentially identical to the straight-through design. One design weakness is that the body at the top of the weir, where the diaphragm seats, is subject to erosion from the fluid and possible eventual leakage.

DIAPHRAGM VALVE APPLICATIONS

Diaphragm valves can be used both as stop valves and for throttling. The straight-through design is more appropriate as a stop valve. When the valve is open, only negligible resistance to flow exists. The weir design is more suitable for throttling. The changes in fluid direction through the valve cause greater resistance to flow and make it less suitable as a stop valve, while at the same time contributing to its ability to regulate flow.

Diaphragm Valves

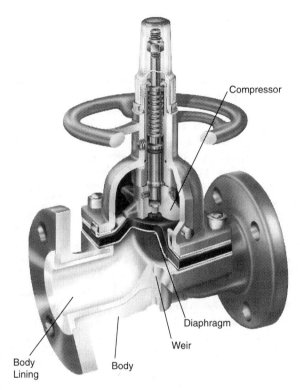

Figure 8-2. Weir diaphragm valve

Whether used as a stop valve or a throttling valve, the following points can be made in favor of the diaphragm valve:

1. It is available in all types of valve materials: bronze, cast iron, carbon, stainless steel, and other corrosion-

resistant alloys. Their design makes diaphragm valves easy to line with abrasion-resistant and chemically resistant non-metallic materials, such as rubber, plastic, and glass.

2. It is a simple design with few parts. Most parts are protected from the corrosive and erosive effect of the fluid. The one part most subject to wear and deterioration, the diaphragm, can be replaced without removing the valve from the pipeline.

3. It is particularly suitable for use with gases and vapors. This is because at least one of the seating parts, the diaphragm, is "soft," making tight shut-off readily attainable.

4. Its smooth, uninterrupted fluid path, with the absence of any pockets that can trap solid material, makes it suitable for use with slurries and viscous fluids. With properly selected diaphragms and body linings, abrasive slurries can be handled.

5. It can be used with highly corrosive fluids such as acids and strong bases. A wide selection of diaphragm, body, and body liner materials resistant to most industrial fluids is available.

6. It is particularly useful in services in which the fluid itself cannot be contaminated. The same materials that are resistant to chemical attack generally do not affect the fluid. This makes the diaphragm valve popular in the food, brewing, pharmaceutical, and other industries in which sanitary conditions must be maintained.

The major weakness of the diaphragm valve, whether the straight-through or weir design, is the diaphragm itself. The materials used for diaphragms have relatively low temperature limits, and they do not possess high tensile strength, even at ambient temperatures. Increasing temperature further reduces the strength of the diaphragm. Because the diaphragm must contain the fluid pressure, diaphragm valves are suitable only for low-pressure, low-temperature applications. Diaphragm valves are capable of handling fluid pressures only up to 200 psi. With elastomer diaphragms, fluid temperatures are limited to about 300°F; with plastic diaphragms, fluid temperatures are limited to about 400°F. Diaphragm weakness also limits the range of sizes of diaphragm valves. They generally are available in sizes from ½ inch up to only 12 inches.

When the weir design diaphragm valve is used for throttling, it exhibits a flow characteristic that falls between the linear and the quick-opening characteristics. Its characteristic is linear (the change in flow is directly proportional to the change in diaphragm travel) up to about 50% valve opening. At this point, the volume of fluid flow through the valve is about 70% of the fully open valve. Beyond this point further opening of the valve produces progressively less change in flow. As a result of this characteristic, if a weir diaphragm valve is to be used for throttling, it should be sized so that the range of flow rate control falls within 70% of the maximum flow that would pass through the valve. (Flow characteristics are discussed in detail in Chapter 11.)

9

Valve Materials

When a valve is called a *bronze valve* or a *steel valve,* reference is being made to the material of construction of the valve shell. The shell of a valve consists of those parts that must contain the fluid pressure. These parts are sometimes called pressure-retaining parts. They are typically the body; the bonnet, cover, or cap; and the bolting used to hold these parts together, if bolting is used in the design of the joint. Because pressure containment is the main function of these parts, strength is the primary consideration in choosing shell materials; however, unless the valve is lined, these parts must also be able to withstand the corrosive and erosive effects of the fluid passing through the valve. Because shell bolting is normally protected from the fluid by gaskets or seals, strength and resistance to atmospheric corrosion are the primary considerations in their selection.

Those parts of a valve inside the shell and in contact with the fluid are referred to as the valve *trim.* Trim parts vary depending on the type and design of the valve, but typically they are the stem or shaft, the flow control element, and the body seating surfaces. It is the function of these parts to perform the operation of the valve, so while adequate strength is necessary, resistance to corrosion and erosion from the fluid and wear resistance are of major importance. This is particularly true for the seating surfaces, because it is the tight, uninterrupted mating of these surfaces that prevents internal leakage when the valve is closed.

In many cases the materials used to make the valve shell do not possess the required properties to be used as trim material. This is the main reason that separate parts such as seat rings and backseat bushings are screwed, pressed, or welded into shell parts rather than being provided as integral machined surfaces. Another reason is that trim materials are generally more expensive than are shell materials. Because of this, it is also common on large valves to make parts, such as wedges and discs, of the same material as the shell, but provide an overlay of trim material on the important seating surfaces.

It is not uncommon to use different materials for different trim parts on the same valve. This is especially true for seating surfaces in which the use of dissimilar metals and metals of different hardnesses lessens the possibility of seizing and galling of the mating surfaces. The most obvious instance of the use of different materials is non-metallic materials being used as body seats in ball and butterfly valves.

A third category of valve parts are those located in the shell, whose function is to prevent external fluid leakage more than to retain pressure. These parts are gaskets and packing. Gaskets form a static seal between shell parts, such as between the body and bonnet of a bolted bonnet valve. They must have the strength to contain fluid pressure, be soft enough to deform and make a tight seal, be as corrosion resistant as is the shell material, be able to withstand high service temperatures, and be able to maintain their properties over time. Packing forms a seal between the shell and the valve stem or shaft. Under normal conditions, when the valve is not being operated, the seal is static; however, during valve operation, packing must also function dynamically. It must be compressible enough to deform and seal tightly against the sides of the packing chamber (also called the *stuffing box)* and the stem or shaft. Packing must be corrosion and temperature resistant, and it must have an inherent low coefficient of friction or contain a lubricant to permit easy stem or shaft movement.

When specifying a valve for purchase, the buyer usually selects shell and trim materials from those available from manufacturers of the type of valve desired. These materials have become standardized, and most valve manufacturers offer the same choices of materials. The shell and trim materials most used in the valve industry are discussed in the subsequent section of this chapter. Bolting, gasket, and packing materials are selected by manufacturers to be compatible with their design, body, and trim materials and, in the case of bolting, to comply with valve standards. The valve buyer usually does not need to specify or select the materials for these parts; however, they are included in this chapter for information.

SHELL MATERIALS

Shell materials can be classified into three broad groups: bronze, cast iron, and steel. The steel group includes the stainless steels. Valves are also made from the more exotic engineering alloys, such as the Inconels® and Hastelloys®; however, these are not as readily available as are the other materials and may need to be custom-made by special order. Each of these three broad groups is considered below.

Bronze and cast iron shell parts are always castings; whereas steel shell parts can be in the cast, forged, or wrought (plate or bar) form. Steel valves in sizes less than

2 inches are usually made of forgings; those larger than 2 inches are usually castings, but they may be forged. Two-inch valves are readily available in either the forged or cast form. Some types of steel valves, such as butterfly valves, are also frequently made from plate steel.

1. Bronze. Bronze is a copper alloy, with tin as the primary alloying element. Lead and zinc are also added in varying proportions to produce the desired properties. The following are the most used valve bronzes:

    - Cast Steam or Valve Bronze, the American Society for Testing and Materials Standard ASTM B61. This is an alloy of high strength and toughness. It is used for shell part castings in higher-pressure classes of bronze valves. In addition to air, water, steam, and oil, it effectively handles many solvents, organic refrigerants, and many other industrial liquids and gases up to 550°F.
    - Cast Composition Bronze, ASTM B62. Also called Ounce Metal or 85-5-5-5, this alloy is used in lower-pressure classes of bronze valves. It combines low cost with a wide range of services, including air, water, steam, oil, and gas up to 450°F.

2. Cast Iron. Cast iron is an alloy of iron, carbon (between 2% and 6.67%), silicon, and manganese. Other elements, such as nickel, are added to produce special forms of cast iron. Cast iron is less susceptible to atmospheric corrosion than is carbon steel. The cast irons most commonly used to make valves are the following:

    - Gray Iron, ASTM A126, Class B. This basic cast iron is easily cast and machined and has good pressure tightness. It is widely used for valves used in water, steam, oil, non-corrosive gas, and some dilute acid services at temperatures from 0°F to 450°F.

- 3% Nickel Gray Iron, ASTM A126, Class B modified. The addition of 3% nickel to Gray Iron improves its grain structure and provides greater corrosion and wear resistance. It is widely used in the petroleum and paper industries.
- Austenitic Gray Iron, ASTM A436, Type 2. Also called Ni-Resist, this highly alloyed (20% nickel) cast iron is widely used as a shell material for valves used in corrosive environments handling caustics, alkalis, ammonia solutions, food products, plastics, and similar services. It is not recommended for steam service.
- Ductile Iron, ASTM A395. This type of cast iron, sometimes known as nodular iron, is unique because the addition of magnesium causes its graphite to be in the nodular or spheroidal form rather than flake form as is found in other cast irons. As a result, strength similar to that of carbon steel and a substantial increase in ductility and toughness over that of gray iron are obtained. Its resistance to atmospheric corrosion is better than that of gray iron and is superior to carbon steel in most cases. It can be used at temperatures down to $-20°F$.

3. Steel. Steels are alloys of iron, carbon (less than 2%), and elements such as manganese and silicon in small amounts. Other elements such as molybdenum, chromium, and nickel are added to impart desired properties of strength, good hardening ability, and corrosion

resistance. The following are those most commonly used to make valves:

- Carbon Steel, ASTM A216 Grade WCB (castings), ASTM A105 (forgings), and ASTM A517 Grade 70 (plate). This is the basic steel used for valve shell parts. Its strength and toughness ensure high resistance to shock, vibration, piping strains, and fire and freezing hazards. It is suitable for saturated or superheated steam; hot or cold water; and hot or cold noncorrosive oils, gases, air, and other fluids. Carbon steel is readily weldable without heat treatment, except in very heavy sections. Temperature limitations are from −20°F minimum to 800°F maximum for continuous service.
- Carbon Steel for Low Temperature Service, ASTM A352 Grade LCB (castings) and ASTM A350 Grade LF2 (forgings). This steel has the same composition and applications as does plain carbon steel, but it is heat-treated to produce similar physical properties at low temperatures. It is recommended for service down to −50°F, but not above 650°F.
- 1¼% Chromium–½% Molybdenum Alloy Steel, ASTM A217 Grade WC6 (castings) and ASTM A182 Grade F11 (forgings). The addition of chromium and molybdenum to carbon steel improves high-temperature strength and resistance to graphization and creep. It is recommended for valves operating at temperatures up to 1,000°F, and it is readily weldable.
- 2¼% Chromium–1% Molybdenum Alloy Steel, ASTM A217 Grade WC9 (castings) and ASTM A182 Grade F22 (forgings). This alloy has higher

resistance to graphitization and creep and higher strength than does 1¼% Chromium–½% Molybdenum steel. It is also recommended for valves operating at temperatures up to 1,000°F. It is weldable but requires pre-heating and heat treatment after welding.

- 5% Chromium Alloy Steel, ASTM A217 Grade C5 (castings) and ASTM A182 Grade F5 (forgings). This alloy can be used in severe services for steam, water, oils, and gases at temperatures up to 1,100°F. It has excellent creep strength and is resistant to corrosion, erosion, graphitization, and scaling. It is particularly recommended for oil refinery service owing to its corrosion resistance to sulfur-bearing crude oils. It is weldable with heat treatment.
- 9% Chromium Alloy Steel, ASTM A217 Grade C12 (castings) and ASTM A182 Grade F9 (forgings). This alloy has excellent strength properties and increased resistance to corrosion. It offers excellent resistance to creep, erosion, and scaling and can be used to 1,050°F. It requires pre-heating before welding and heat treatment after welding.
- 18-8 Austenitic Stainless Steel, ASTM A351 Grade CF-8 (castings) and ASTM A182 Grade F304 (forgings). This 18% chromium, 8% nickel stainless steel is suitable for oxidizing and very corrosive fluids. It is also recommended for extremely corrosive oil and liquid oxygen service and is particularly suited for nitric acid service. It is frequently used above 1,000°F and below −150°F. It is weldable without heat treatment.
- 18-8-3M Austenitic Stainless Steel, ASTM A351 Grade CF-8M (castings) and ASTM A182 Grade

F316 (forgings). The addition of 3% molybdenum to 18-8 stainless steel results in improved resistance to corrosion at high and low temperatures for oil, dilute hydrochloric acid and other acids, brine, and other process fluids. It is weldable.
- Alloy 20, ASTM A351 Grade CN-7M (castings), ASTM B462 UNS N08020 (forgings), and ASTM B473 UNS N08020 (wrought). This 20% chromium, 29% nickel alloy has excellent resistance to sulfuric acid over a wide range of concentrations. It also has good resistance to dilute hydrochloric acid and is used extensively in the manufacture of high-octane gasoline, solvents, and other process fluids. It can be satisfactorily welded.

4. Bolting. Body–bonnet and body–cap bolting materials are selected by valve manufacturers to be compatible with the other shell materials and to satisfy valve standards. The following are the typical bolting materials used on valves:

- Free Machining Brass Rod, ASTM B16. Also known as commercial brass, it is used to make nuts for bronze valves.
- Carbon Steel Bolts and Studs, 60,000 psi Tensile Strength, ASTM A307. Grade A is commonly used on bronze valves. Grade B, which has tighter controls on hardness and tensile strength, is used on cast iron valves and may be used on steel valves, provided its 500°F maximum temperature limitations are met.
- Carbon and Alloy Steel Nuts, ASTM A563 Grade A. These nuts are used with A307 bolts and studs on cast iron and steel valves.

- Alloy Steel and Stainless Steel Bolting Materials for High Temperature Service, ASTM A193. Grade B7, a chromium–molybdenum steel suitable for temperatures up to 1,000°F, is the standard material for bolts and studs used on steel valves. Other grades such as B16M, B8M, and B8M are used on alloy and stainless steel valves and to meet unusual temperature or corrosion conditions.
- Carbon and Alloy Steel Nuts for High Pressure and High Temperature Service, ASTM A194. Grade 2H is the standard grade nut for use with B7 studs and bolts. Other grades such as 3M, 8M, and 8M are used with bolts and studs of comparable material.

TRIM MATERIALS

Trim materials can be classified into two broad groups, metallic and non-metallic. The metallic group includes the copper alloys, the stainless steels, and some of the exotic engineering alloys. The non-metallic group includes plastics and elastomers.

Some generalizations can be made about trim materials. Bronze valves generally have brass and bronze trim. Cast iron valves have mostly bronze trim (called IBBM, iron body, bronze mounted), or iron trim (called all-iron). Carbon and alloyed steel valves are available in a wide range of trim materials, including bronze and stainless steel. The trim material for stainless steel valves is generally the same as the shell material.

1. Metallic Materials. Where these materials are also used as shell material, refer to that section for a description.

- Free Machining Brass Rod, ASTM B16. Commonly known as commercial brass, this metal is used for stems on cast iron and cast steel valves and balls and stems on bronze ball valves.
- Cast Steam Bronze, ASTM B61. This bronze is used for wedges, discs, and seat rings in ductile iron valves.
- Cast Composition Bronze, ASTM B62. This bronze is used for discs and seat rings in cast iron and cast steel valves.
- Aluminum Bronze, ASTM B148. This 10% aluminum, 5% iron, copper alloy is suitable for a variety of medium-corrosive service conditions. It is used for discs in ductile iron and steel butterfly valves.
- Wrought Copper–Silicon Bronze, ASTM B371. This special copper–zinc–silicon bronze is the standard bronze valve stem material. It has outstanding corrosion resistance, high strength, and excellent wearing properties.
- Cast Copper–Nickel Bronze, ASTM B584 C97600. This 20% nickel alloy is an excellent non-galling material that possesses high strength with mild corrosion resistance. Its hardness makes it suitable for seating surfaces in bronze valves.
- Cold Drawn Steel Bar, ASTM A108, Grade 1018. This low carbon steel is strong enough to be used as stems in all-iron valves. It is either chemically treated or plated to inhibit rust.
- Austenitic Gray Iron, ASTM A436, Type 2. This nickel alloyed cast iron is used for wedges and discs on 3% Nickel Gray Iron valves.

Valve Materials

- 13% Chromium Martenistic Stainless Steel (CR 13), ASTM A217 Grade CA15 (cast), ASTM A182 Grade F6a (forged), ASTM A276 Type 410 (wrought), and American Welding Society AWS A5.9, ER410 (weld rod). This stainless steel is an excellent valve trim material for a wide range of service conditions in which severe corrosion resistance is not required. It is recommended for oil, oil vapor, and mild corrosives up to 1,000°F. Where one CR 13 component bears against another, such as body and wedge seating surfaces, a 50 Brinnell hardness differential between the surfaces is necessary to prevent galling. It is used for stems, wedges, and body seats on steel gate, globe, and check valves; shafts on ductile iron and steel butterfly valves; and stems on plug valves.
- 18-8 Austenitic Stainless Steel, ASTM A352 Grade CF-8 (cast), ASTM A182 Grade F304 (forged), ASTM A276 Type 304 (wrought), and AWS A5.9 ER308 (weld rod). This stainless steel is used for stems, wedges, discs, and body seats on steel gate, globe, and check valves; discs on steel butterfly valves; and stems on plug valves.
- 18-8-3M Austenitic Stainless Steel, ASTM A351 Grade CF-8M (cast), ASTM A182 Grade F316 (forged), ASTM A276 Type 316 (wrought), and AWS A5.9 ER 316 (weld rod). This stainless steel is used for stems, wedges, discs, and body seating surfaces on steel gate, globe, and check valves; discs, shafts, and seating surfaces on butterfly valves; balls, stems, and seat rings on ball valves; stems and seat rings on 3% nickel gray iron valves; and all trim parts on austenitic gray iron valves.

- 17-4PH Age Hardening Stainless and Heat Resisting Steel Bar, ASTM A564 Type 630. This 17% chromium, 4% nickel precipitation hardening martensitic stainless steel retains its strength and corrosion resistance at high temperatures. It is used for shafts in butterfly and ball valves, seat rings in ball valves, and stems in steel gate and globe valves in which strength at high temperatures is required.
- Monel® Nickel–Copper Alloy, ASTM A494 Grade M-25S (cast), ASTM B564 UNS N04400 (forged), and ASTM B164 UNS N04400 (wrought). This 65% nickel, 30% copper alloy has excellent resistance to corrosion from many acids, caustic soda, alkalis, food products, organic substances, brine, and salt solutions. It is especially suitable for hydrofluoric acid and chlorine service. Monel® is used for stems, wedges, discs, and body seating surfaces in steel gate, globe, and check valves; discs and shafts in butterfly valves; and balls and shafts in ball valves.
- Alloy 20, ASTM A351 Grade CN-7M (cast) and ASTM B473 UNS N08020 (wrought). This alloy is used for stems, wedges, discs, and body seats in gate, globe, and check valves; discs and shafts in butterfly valves; and balls and stems in ball valves.
- Hastelloy C®, ASTM A494 Grade CW-12MW (cast) and ASTM B574 UNS N10276 (wrought). This nickel-based, 16% chromium, 16% molybdenum alloy is used in severe service conditions that usually involve acids at high temperatures. It is resistant to strong oxidizers and has high tem-

perature strength. It is used for stems, wedges, discs, and body seats in gate, globe, and check valves.
- CoCr-A Hardfacing, AWS A5.13 ERCoCr-A (weld rod). This cobalt-based, 30% chromium, 5% tungsten alloy includes trademarked materials such as Stellite® 6. It maintains its wear resistance, corrosion resistance, and hardness under severe high-temperature (up to 1,500°F), high-pressure conditions. The hardness of as-deposited hardfaced material measures in the 400 to 450 Brinnell hardness range. It is used as a welded overlay to form seating surfaces on gate, globe, and check valves and on seat rings on metal-seated ball valves.

Most manufacturers of cast steel gate, globe, and check valves offer different combinations of metal trim materials. These are shown in Table 9-1.

Table 9-1. Combinations of metal trim materials

Trim*	Wedge/Disc**	Seat(s)	Stem
1	CR 13	CR 13	CR 13
5	Hardfaced	Hardfaced	CR 13
8	CR 13	Hardfaced	CR 13
9	Monel	Monel	Monel
10	18-8-3M	18-8-3M	18-8-3M
12	18-8-3M	Hardfaced	18-8-3M
13	Alloy 20	Alloy 20	Alloy 20
N/A	Bronze	Bronze	Bronze

*From American Petroleum Institute Standard API 600, Table 3.
**Solid or facing.

Trim Number 1 is recommended for oil and oil vapor service at temperatures to 1,000°F. It is also used for steam, water, air, and gas service for temperatures to 850°F. This trim has generally been supplanted by Trim Number 8, which has become a valve industry standard, for the same applications. Trim Number 5 is recommended for oil and oil vapor service up to 1,000°F. It is particularly suitable for high-pressure steam and water service up to 1,200°F. Trim Number 9 has wide application in handling alkalis, sea water, brine, organic substances, nonoxidizing solutions, and many air-free acids up to 450°F. Numbers 10 and 12 trims are recommended for corrosive oil service and for acetic, hydrochloric, and phosphoric acids up to 850°F. Number 13 trim is recommended for all concentrations of sulfuric acid and for dilute hydrochloric acid. Bronze trim is used primarily for saltwater and freshwater service up to 450°F.

2. Non-metallic Materials. Non-metallic trim materials fall into two categories, plastics and elastomers. Plastics are organic polymeric materials (they are made of giant organic molecules) that possess a degree of structural rigidity—an elastic modulus on the order of 15,000 psi. They are characterized by high strength-to-density ratios, excellent thermal and electrical insulation properties, and good resistance to acids, alkalis, and solvents. Plastics can be either thermoplastic or thermosetting. The thermoplastic type can be re-softened to its original condition by heat; the thermosetting type cannot. Plastics are generally more sensitive to high temperatures than are metals. Reinforced plastics are made by adding fibrous materials, such as glass fibers, to alter the physical properties of the plastic.

Valve Materials

Elastomers are also organic polymeric thermosetting materials, but they have properties similar to vulcanized natural rubber. They have the ability to be stretched to at least twice their original length, and they retract very rapidly to their original length when released.

The following are the non-metallic materials most commonly used as valve trim:

- Buna-N is an elastomer also known as nitrile rubber and NBR and is a copolymer of butadiene and acrylonitrile. It has excellent resistance to abrasion, tearing, and compression set. It is a good, general-purpose material suitable for use with air, water, and alcohols. It is recommended for use with mineral oils, gasoline, and butane and natural gases. Buna-N is not recommended for concentrated acids, solvents, and strong oxidizing agents. Its service temperature ranges from $-20°F$ to $180°F$. It is used for butterfly valve liners and seats and for diaphragm valve diaphragms.
- EPT is an elastomer also known as EPDM, EP, EPR rubber, Nordel®, and ethylene-propylene. It is a compound of ethylene, propylene, and usually a third monomer. It has very good resilience and good abrasion resistance. It is recommended for use with air, freshwater, saltwater, alcohols, hydrogen sulfide, and sulfur dioxide. EPT is not recommended for use with mineral oils, gasoline, gases, and solvents. Its service temperature ranges from $-40°F$ to $300°F$. It is used for butterfly valve liners and seats and for diaphragm valve diaphragms.
- Hypalon® is a thermoplastic also known as CSM and is a chloro-sulfonated polyethylene com-

pound. It has excellent tear and abrasion resistance but poor compression-set resistance. Hypalon® is recommended for use with acids, alkalis, alcohols, oxidizing fluids, and aqueous salt solutions. It is not recommended for water, solvents, mineral oils, or gasoline. Its service temperature ranges from 0°F to 225°F. It is used for butterfly valve liners and diaphragm valve diaphragms.
- Natural rubber has excellent tear and abrasion resistance and very good resilience. It is recommended for use with freshwater and saltwater, dilute acids and alkalis, and some sulfate solutions. It is not recommended for use with mineral oils, gasoline, or solvents. Its service temperature ranges from −60°F to 160°F. It is used for butterfly valve linings and for diaphragm valve linings and diaphragms.
- Neoprene is an elastomer also known as chloroprene rubber and is a polymer of chloroprene and chlorobutadiene. It has excellent abrasion resistance and compression-set resistance and very good resilience. The addition of carbon black increases its tear resistance. Neoprene is recommended for use with air, water, alcohols, freons, and natural gases. It is not recommended for use with acids, solvents, or petroleum products. Its service temperature ranges from −40°F to 200°F. It is used for butterfly valve liners and for diaphragm valve linings and diaphragms.
- PEEK is a thermoplastic also known as polyether-etherketone and is an aromatic polyketone. It is best-suited for high temperatures and

up to 475 psi service with steam. It is not suitable for use with strong oxidizing acids, such as sulfuric acid, and bases at high concentrations. Its service temperature ranges from 0°F to 550°F. It is used for ball and butterfly valve seats.

- Teflon® is thermoplastic also known as tetrafluoroethylene (TFE), Virgin TFE, and PTFE (poly-TFE). It is a polymer of carbon and fluorine. It is abrasion resistant, relatively soft but tough, and has a very low coefficient of friction. Teflon® can be used with air, water, brine, acids, alkalis, solvents, mineral oils, gasoline, and saturated steam up to 125 psig. It is not recommended for fluorine gas, freons, concentrated sulfuric acid, or high-pressure steam. Its service temperature ranges from −100°F to 400°F. Glass fibers are added to Teflon® to produce reinforced, or filled, Teflon® (RTFE). It retains almost all the chemical resistance properties of Teflon®, extends the service temperature range up to 500°F, improves abrasion resistance, and is suitable for saturated steam service up to 150 psig. Teflon® and reinforced Teflon® are used for ball valve seats, butterfly valve liners and seats, diaphragm valve diaphragms, and plug valve sleeves.

- Tefzel® is a thermoplastic also known as ETFE and is a modified copolymer of ethylene and tetrafluoroethylene. It is tough and especially resistant to gamma radiation. It is recommended for strong and weak acids and for gases, and it has no known solvent below 392°F. Its service temperature ranges from −50°F to 360°F. It is used for butterfly valve seats and diaphragm valve linings.

- Viton® is an elastomer also known as Fluorel®, FKM, and fluorocarbon. It is a fluorine-containing hydrocarbon polymer. It has good abrasion and tear resistance and resilience, but poor compression-set resistance. It is recommended for use with water, acids, solvents, mineral oils, and gasoline; it is not recommended for use with ammonia and steam. Its service temperature ranges from 0°F to 400°F. It is used for butterfly valve seats and liners and for diaphragm valve diaphragms.

To assist valve buyers in specifying appropriate non-metallic material, most valve manufacturers usually provide application information (called material selection or chemical resistance guides) showing the compatibility of the different plastics and elastomers they offer with specific corrosive media.

GASKETS AND PACKING

1. Gaskets. The common types of gaskets used in valve construction are flat, spiral-wound, and ring joint gaskets. Flat gaskets are used on low-pressure valves and can be made of either non-metallic materials or metal. Spiral-wound and ring joint gaskets are used on medium- and high-pressure valves. Spiral-wound gaskets are composed of alternate plies of preformed metal bands and non-metallic filler; ring joint rings are metal.

   Non-metallic flat gaskets are cut from sheets of material that can be classified into three categories:

   - Compressed sheets comprised of a mixture of fibers, fillers, and an elastomeric binder. In the

past, natural asbestos was the standard fiber material; however, in recognition of the health hazards associated with asbestos, synthetic fibers such as glass, aramid, and carbon have come into use. Synthetic fiber gaskets generally do not possess all of the desirable properties of asbestos fiber gaskets.
- Homogeneous sheets of Teflon® and reinforced Teflon®. Teflon® gaskets have low-service-temperature, high-corrosion applications.
- Laminated sheets of flexible graphite foil made from natural flake graphite. Metal reinforcing inserts are sometimes included for added strength. These gaskets are highly temperature and corrosion resistant.

Metal flat gaskets can have smooth or corrugated surfaces. These gaskets are commonly made of iron or soft steel, but the stainless steels are also used.

A wide range of metal and filler combinations are available in spiral-wound gaskets to match corrosion resistance and sealing requirements. The metals include carbon and stainless steels, Monel, and the Hastelloys®. Fillers include asbestos, TFE, and flexible graphite. The combination of 316SS with graphite filler is very common. These gaskets are typically manufactured with carbon steel centering rings.

Ring joint rings have oval- or octagonal-shaped cross-sections. They are made of metals comparable with valve shell materials: carbon steel, low alloy steel (e.g., 5% Cr–1% Mo), and stainless steel (e.g., 304 and 410). Rings usually are softer than are the flanges with which they are used.

2. Packings. The basic packing materials and construction commonly used for valve stem and shaft packing are the following:

- Braided. Strands of yarn are wound so that the strands pass over and under strands running in the same direction, or criss-cross from the surface diagonally through the body of the packing (called an *interbraid*). In the first type the braid may be over a core, which may extruded, twisted, or wrapped. In the past, long-fiber asbestos was the standard yarn material. Today, carbon fiber and synthetic yarns are used almost exclusively. Inconel wire is usually twisted with or inserted into the yarns for added strength. Flake graphite or molybdenum disulfide are used as lubricants and sacrificial metals such as zinc are added to inhibit corrosion.
- Molded plastic. Formed rings of solid plastic, typically TFE, are used in low-temperature, high-corrosion applications. They can be molded into many cross-section shapes, for example, square and chevron.
- Die-formed, flexible graphite. Flexible graphite foil is slit into ribbons and corrugated. The ribbons are then wound in dies and compressed into dense rings. A suitable agent is included to inhibit corrosion. As with molded plastic rings, these rings can be formed into different cross-section shapes. When used with braided top and bottom rings that act as wipers and limit extrusion, these rings are very effective packing sets for use against gas and vapor leaks.

10

Sizes, Classes, and Ratings

VALVE SIZES

The size of a valve is defined by the size of the pipe to which it can be connected. That is, a 2-inch valve has ends that connect with 2-inch pipe, an 8-inch valve has ends that connect with 8-inch pipe, and so on. Of course, the valve ends and pipe ends must be compatible.

Valves are manufactured in all standard pipe sizes. However, for the less common pipe sizes—3½, 22, 26, and 28 inches—and pipe sizes larger than 30 inches, valves are not usually listed in catalogs or made for manufacturers' or suppliers' stock. They must be custom-made by special order. Other sizes—⅛, ⅜, 1¼, and 5 inches—are listed in manufacturers' catalogs but are not always available from stock.

There are additional limitations on available valve sizes, depending on the valve type, shell material, and pressure class. The ranges of valve sizes for the different valve types have been covered in the chapters in which the different types are discussed and will not be repeated here. For the commonly used shell materials the following sizes are generally available:

1. Bronze—⅛ inch to 3 inches.
2. Cast iron—2 inches to 30 inches.

3. Forged steel—Primarily ¼ inch to 3 inches; however, some high-pressure gate, globe, and check valves in sizes up to 24 inches are made from forgings.
4. Cast steel—2 inches to 30 inches.
5. Fabricated steel—can be made in any size, but these valves are usually made in sizes larger than cast steel valves for cost reasons.

For the less commonly used shell materials, such as ductile iron, Alloy 20, Monel®, and so on, the full range of sizes listed previously is not readily available, particularly in the large sizes. Also, pressure class (see below) can limit the availability of valve sizes. Generally, the higher the pressure class, the smaller the range of sizes.

There is another aspect to the size of a valve, its length, which is the space the valve takes up in the pipeline. The length of a valve is expressed by its end-to-end dimension or, in the case of flanged end valves, its face-to-face dimension. For a given valve size the end-to-end or face-to-face dimension varies with the type, design, and pressure class of the valve. To facilitate piping design, dimensional standards for valve length have been established. For instance, the American Society of Mechanical Engineers Standard ASME B16.10 specifies the end-to-end and face-to-face dimensions for different types and designs of butt-weld end and flanged end valves. Manufacturers rarely deviate from these standards.

Valve height is also a size consideration. Height is determined primarily by valve type. The translating stem gate, globe, and diaphragm valves are higher than are stemless check valves and quarter-turn butterfly, ball, and plug valves, which do not have bonnets. Also, the gate valve, with its longer stroke, is higher than the same size globe or diaphragm valve. Dimensional standards do not exist for valve height, but information can be found in manufacturers' catalogs. This information is important when designing piping systems to prevent the valve from interfering with adjacent piping and equipment.

VALVE CLASSES

Most valves are manufactured in pressure classes. These classes approximately indicate the pressure-containing capability of a valve. The higher the class number, the greater the pressure-containing capability. Bronze valves are made in Classes 125, 150, 200, and 300; cast-iron valves are made in Classes 125, 150 and 250; and steel valves are manufactured in Classes 150, 300, 400, 600, 800, 900, 1500, 2500, and 4500. For the steel valves, Class 400 is not readily available, and Class 800 is available only in valves with forged shell parts

Pressure classes are frequently referred to as *pounds.* For instance, a Class 600 valve is often call a *600 pound valve.* This can be misleading, because a Class 600 valve can contain an internal pressure of much more than 600 psi under some conditions, but less than 600 psi under others.

Different designs of a type of valve may be used for different pressure classes. For example, a pressure-seal, body–bonnet joint design may be used instead of a bolted-bonnet design for a valve in a high-pressure class. The important difference between classes, however, is the wall thickness of the valve shell. Standards such as ASME B16.34 and API 600 specify the required shell wall thicknesses for different valve pressure classes: the higher the pressure class, the thicker the wall. For example, ASME B16.34 requires a 4-inch valve in Class 150 to have a 0.25-inch-thick shell; whereas a Class 300 valve needs a 0.31-inch-thick shell, and a Class 1500 needs a 0.75-inch-thick shell.

VALVE RATINGS

The allowable stress on a metal decreases as the temperature of the metal increases. Therefore, the actual pressure-containing capability of a valve is determined by the pressure-temperature rating of the shell material of the valve. (A valve shell is assumed to have the same temperature as the fluid passing through it.)

Figure 10-1 shows Tables 2-1.1 and 2-1.1A from ASME B16.34. These tables give the pressure-temperature ratings for valves made of cast, forged, and wrought carbon steel with shells having the minimum wall thicknesses required by that standard. Table 2-1.1A shows that the higher the valve's shell temperature, the lower its pressure-containing capability (see "Working Pressure" in Figure 10-1, Table 2-1.1A). The pressure capability of a carbon steel shell for a particular class is much higher than the class designation at ambient temperature (−20°F to 100°F), and it decreases as the temperature of the valve increases until it is below the class designation. For example, a Class 150 carbon steel valve can be used at 285 psi at ambient temperature but should be used at only 80 psi at 800°F. Consequently, if a piping system is to contain a fluid at a pressure of 150 psi and a temperature of 600°F, Class 150 valves cannot be used. At 600°F, they are rated for only 140 psi. The valves in that piping system must be Class 300 valves, which are rated for 550 psi at 600°F. Another example would be a valve used for steam service at 800 psi and 750°F. Referring to Figure 10-1, it can be seen that a Class 600 valve must be used. At 750°F, a Class 600 valve has a working pressure of 1,010 psi; whereas a Class 400 valve has a working pressure of only 670 psi.

Tables comparable to Table 2-1.1A in Figure 10-1 for alloy and stainless steels can also be found in ASME B16.34. Note that the pressure-temperature ratings of

TABLE 2-1.1
RATINGS FOR GROUP 1.1 MATERIALS

A 105(a)	A 515 70(a)	A 675 70	A 672 B70(a)
A 216 WCB(a)	A 516 70(a)	A 696 Gr. O	A 672 C70(a)
A 350 LF2(d)	A 537 Cl.1(d)		

NOTES:
(a) Permissible, but not recommended for prolonged usage above about 800°F.
(d) Not to be used over 650°F.

TABLE 2-1.1A STANDARD CLASS

Temperature, °F	Working Pressure by Classes, psig							
	150	300	400	600	900	1500	2500	4500
−20 to 100	285	740	990	1,480	2,220	3,705	6,170	11,110
200	260	675	900	1,350	2,025	3,375	5,625	10,120
300	230	655	875	1,315	1,970	3,280	5,470	9,845
400	200	635	845	1,270	1,900	3,170	5,260	9,505
500	170	600	800	1,200	1,795	2,995	4,990	8,980
600	140	550	730	1,095	1,640	2,735	4,560	8,210
650	125	535	715	1,075	1,610	2,685	4,475	8,055
700	110	535	710	1,065	1,600	2,665	4,440	7,990
750	95	505	670	1,010	1,510	2,520	4,200	7,560
800	80	410	550	825	1,235	2,060	3,430	6,170
850	65	270	355	535	805	1,340	2,230	4,010
900	50	170	230	345	515	860	1,430	2,570
950	35	105	140	205	310	515	860	1,545
1000	20	50	70	105	155	260	430	770

Figure 10-1. Pressure-temperature ratings of carbon steel

Class 800 valves can be found from these tables by interpolating between Classes 600 and 900. Pressure-temperature tables for bronze and cast iron sheet materials can be found in the Manufacturers Standardization Society of the Valve and Fittings Industry Standards MSS SP-70, MSS SP-71, MSS SP-80, and MSS SP-85. Many manufacturers' catalogs also contain this information.

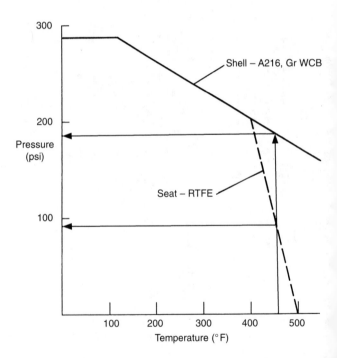

Figure 10-2. Pressure-temperature rating of a Class 150 ball valve

Sizes, Classes, and Ratings

Valve types with non-metallic seating parts—the ball, butterfly, and diaphragm valve—are subject to pressure-temperature limitations in addition to that imposed by the metal shell materials. With the ball and butterfly valves, the seats also have a pressure-temperature rating with respect to the differential pressure across them with the valve in the closed position. This is due to pressure-induced deformation of the non-metallic seat rings at elevated temperature. Seat ratings are usually shown in a graphical form, rather than the tabular form used for shell materials. A graph similar to that which would be found in manufacturers' catalogs for a Class 150 carbon steel ball valve is shown in Figure 10-2.

In Figure 10-2 the solid line shows the shell material rating, which is drawn from the data in Figure 10-1. The dashed line shows the rating for reinforced TFE seats. It can be seen that up to 400°F, the seat rating is the same as the shell material; however, above 400°F the seat rating comes into play. For example, at 450°F the shell is capable of withstanding 185 psi, but the seats can stand only 90 psi differential pressure. This means that with a fluid at 450°F a closed valve can be pressurized up to 185 psi without fear of shell failure, but it may not close tightly. To ensure tight closure, this valve should only be used when the inlet pressure does not exceed 90 psi.

In the diaphragm valve, the pressure-temperature rating of the valve is the same as that of the diaphragm. Manufacturers typically present such diaphragm and valve rating information in a graphical form. A graph similar to one that would be found in manufacturers' catalogs is illustrated in Figure 10-3. It shows a family of curves, each of which has a constant pressure rating to 120°F and then a uniformly decreasing pressure as temperature increases. The allowable pressures at low temperatures are lower and their

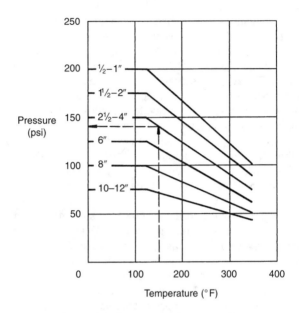

Figure 10-3. Pressure-temperature ratings
of a diaphragm valve

rate of decline more rapid than for metal shells, reflecting the lower tensile strength and greater temperature sensitivity of the non-metallic diaphragm materials. The different curves for different valve sizes are the result of the increasing loads on the diaphragms as the valves increase in size. Because no provision is made in these rating curves for the different diaphragm materials, it must be assumed that all the materials have approximately the same strength and temperature characteristics.

To determine a valve's pressure-temperature rating,

look at the temperature axis in Figure 10-3. Find the intersection point of the vertical temperature line with the curve for the valve in question, and read the allowable working pressure from the pressure axis. Figure 10-3 is marked to show an allowable pressure of 140 psi for a 4-inch valve carrying fluid at 150°F. When determining pressure-temperature ratings, the maximum temperature limitations of the diaphragm materials must be recognized. Some manufacturers show these limits as vertical lines on the graph.

It should be remembered that for a diaphragm valve, diaphragm failure does not constitute valve failure in the same sense that shell failure does for the other types of valves. When a diaphragm fails the valve is inoperable and cannot perform its control function; however, the bonnet maintains the integrity of the valve shell and prevents drastic leakage to the surrounding environment.

VALVE MARKINGS

Valve standards require certain information about a valve to be marked on its body, identification plate, or both. The required information typically includes the manufacturer, shell material identification, size, and pressure class. The way in which the pressure-class information is marked on the valve varies, depending on the shell material and intended service of the valve. On valves made of steel (including alloys and stainless steel) only the Class numbers (e.g., 150, 300) are shown. On valves made of bronze and cast iron, usually there are two numbers. One number is identified by the letters S, SP, or SWP, for steam, steam pressure, or steam working pressure, respectively. This number is the valve class and also shows the maximum allowable working pressure of the valve when used with steam, provided the steam temperature does not exceed the maximum temperature for the shell material. The other number is identified by the letters WOG, designating Water, Oil, Gas. This number is the allowable working pressure of the valve when used at ambient temperatures and is also frequently referred to as the Cold Working Pressure (CWP). The letters WOG or CWP may also be marked on a valve that is not intended to be used at elevated temperatures, which may be allowed by the shell material but not the non-metallic trim materials.

11

Fluid Flow Through Valves

When a valve is first cracked open, fluid begins flowing through it. Further movement of its flow control element allows more fluid to flow, until the valve is completely open and the maximum fluid flow rate is achieved. The flow rate through the valve when it is completely open depends on the resistance to flow of the entire pipeline in which the valve is installed, and the valve contributes to this total line resistance. Consequently, there are two distinct aspects to fluid flow through valves. First, there is the effect of the position of the flow control element on the flow rate through the valve when it is partially open. Second, there is the resistance to fluid flow of the valve when it is completely open.

VALVE FLOW CHARACTERISTICS

The flow characteristic of a valve is the relationship between the position of its flow control element and the rate of fluid flow through the valve. It is due to the changing shape and area of the opening that the flow control element produces as it moves through its travel. Therefore, the flow characteristic is determined by the type and design of the valve.

Inherent valve flow characteristics are determined by flow tests in which the decrease in pressure across the valve is kept constant. The test results are plotted as curves on graphs in which flow control element position and flow rate are shown as percentages of their maximums. Figure 11-1 is a plot of the three common valve flow characteristics: quick-opening, linear, and equal percentage. The quick-opening characteristic produces large changes in flow rate at the start of flow control element travel, and then progressively smaller changes until the valve is completely open. The linear characteristic produces changes in flow rate that are directly proportional to flow control element position for the full range of flow con-

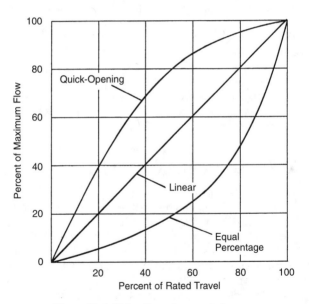

Figure 11-1. Valve flow characteristic curves

trol element travel. With the equal percentage characteristics, equal increments of flow control element travel produce equal percentage changes in flow rate. Stated another way: the change in flow rate is proportional to the flow rate at the start of the change.

To illustrate the equal percentage characteristic, consider the equal percentage flow characteristics curve in Figure 11-1. The curve shows that changing the flow control element position from 40% open to 60% open (a 20% increment) produces a change in flow rate from 12% of maximum to 24% of maximum—a 100% increase. Changing the flow control element position from 60% open to 80% open (also a 20% increment) produces a change in flow rate from 24% of maximum to 48% of maximum— again, a 100% increase.

The flow characteristic exhibited by a valve when it is in a pipeline (known as the *installed flow characteristic*) deviates somewhat from the inherent characteristics shown, because decrease in pressure across the valve varies with flow and other changes in the piping system. In general, the closer the valve is to being fully open, the less effect its flow control element position has on flow rate.

Valves exhibiting the linear flow characteristic and the equal percentage flow characteristic are preferred for use in flow regulation (called *throttling*) applications. With these characteristics, reasonably accurate predictions of flow rate can be made from flow control element position overall, or at least for most of the entire range of flow control element travel. Of the valves used in throttling service, the globe valve has a linear characteristic, the lined butterfly valve has an equal percentage characteristic, and high-performance butterfly and diaphragm valves have characteristics that fall between the linear and equal percentage characteristics.

The quick-opening flow characteristic is found on valves used primarily as stop valves, such as gate and ball valves.

VALVE FLOW RESISTANCE

At the beginning of this chapter it was noted that the resistance to flow of a valve is one component of the total resistance to flow in a pipeline. In Chapters 2–8, references have been made to the resistance to flow of the different types and designs of valves. In this section we show how such resistance to flow can be quantified and make some comparisons of flow resistance between different valve types.

The fluid flowing through a piping system loses energy because of following effects:

1. Pipe friction, which is a function of the inner surface roughness of the pipe, pipe size, and fluid properties.
2. Changes in flow direction caused by elbows, returns, and tees.
3. Changes in pipe cross-section caused by reducers.
4. Obstructions to flow caused by valves, strainers, and so on.

Loss of energy is manifested by a decrease in fluid pressure. Fluid pressure is usually expressed as pounds per square inch (psi) or as feet of "head." A foot of head exerts a pressure equal to that produced by the weight of a column of the flowing fluid one foot high. Pressure and head are related by the equation

$$h = \frac{144p}{\rho} \qquad (1)$$

Fluid Flow Through Valves

where: h = head (ft)
p = pressure (psi)
ρ = fluid weight density (lb/ft^3)

Piping designers often use head to express pressure because it is easy to incorporate the changes in elevation that a typical piping system goes through.

Using head, decrease in fluid pressure caused by pipe friction can be expressed by the equation

$$h_L = f\left(\frac{L}{d}\right)\frac{V^2}{2g} \qquad (2)$$

where: f = friction factor (dimensionless)
L = length of pipe (ft)
d = pipe diameter (ft)
v = fluid velocity (ft/s)
g = gravitational constant (32.2 ft/s^2)

Research has shown that the friction factor is a function of the inner surface roughness and diameter of the pipe, and of the fluid velocity, density, and viscosity. Graphs developed to show the relationships among these variables are used to determine the friction factor in a particular case.

Equation (2) is valid for all fluids, both incompressible (e.g., water and oil) and compressible (e.g., steam and gases). Because of changes in the fluid density of compressible fluids, however, when the change in the decrease of pressure exceeds 40% of the inlet pressure, more complex equations incorporating fluid density corrections are required.

Also using head, decrease in pressure caused by valves, obstructions, changes in flow direction, and changes in pipe cross-section can be expressed by the equation

$$h_L = K \frac{V^2}{2_g} \qquad (3)$$

where K is a dimensionless resistance coefficient. The Valve Resistance Coefficient is a function of valve geometry and therefore is constant for all conditions of flow. The Resistance Coefficient varies somewhat with the size of the valve owing to unavoidable changes in geometry. However, because that variation is much less than the variation between valve types and designs, for comparison purposes, the resistance coefficient will be considered to be constant for all sizes.

Comparing Equations (2) and (3), it can be seen that

$$K = f\left(\frac{L}{d}\right) \qquad (4)$$

and

$$L = \frac{Kd}{f} . \qquad (5)$$

Consequently, by using an appropriate friction factor, Equation (5) can be used to calculate the length of straight pipe of the appropriate diameter that would be equivalent to any valve for which a resistance coefficient has been determined.

It has been found convenient for some types of valves, particularly those used in throttling applications, to express flow characteristics in terms of the Flow Coefficient C_v. The flow coefficient of a valve is defined as the volume of water at 60°F, in gallons per minute (gpm), that will flow through the valve at a decrease in pressure of 1 psi across the valve. The volume of flow for other fluids and other pressure differentials can be found by using the equation

$$Q = C_v \sqrt{\frac{\Delta P}{\delta}} \qquad (6)$$

where: Q = flow (gpm)
ΔP = pressure drop (psi)
δ = specific gravity of fluid (dimensionless)

By using this equation the Flow Coefficient can also be used to calculate the decrease in pressure across the valve for different volumes of flow and for fluids other than water.

By using Equations (1), (3), and (6) and the relationship

$$Q = AV \qquad (7)$$

where A is the flow passage area in square feet, it can be shown that valve Flow Coefficients and Resistance Coefficients are related by the equations

$$C_v = \frac{4{,}300 d^2}{\sqrt{K}} \qquad (8)$$

$$K = \left(\frac{4{,}300 d^2}{C_v}\right)^2. \qquad (9)$$

From these equations we see that the larger the Resistance Coefficient, the smaller the Flow Coefficient, and vice versa.

Valve manufacturers conduct flow resistance tests on their products and publish the results in their catalogs. The results are presented in one of the forms discussed previously: Resistance Coefficient, equivalent pipe length, or Flow Coefficient. These data are used by piping designers when determining the best size for a piping system, and by others in cases in which the flow rate or the decrease in pressure across a valve is a consideration in selecting the appropriate valve for a particular application. By using Equations (4), (5), (8), and (9), data from different manufacturers in the different forms can be converted to a single form to enable comparison.

Typical values of Resistance Coefficients, equivalent lengths, and Flow Coefficients for some of the different valve types are listed in Tables 11-1, 11-2, and 11-3. It should be remembered when examining these data that they are based on fully open valves.

Table 11-1. Resistance coefficient K

Type	Gate	Globe	Swing Check	Ball*	Butterfly**
K	0.2	10	1.4	0.1	1.3

*Full port.
**Concentric disc and seat.

Table 11-2. Equivalent feet of pipe*

Size/Type (inches)	Gate	Globe	Swing Check	Ball	Butterfly
1	0.6	30	4.2	0.3	—
2	1.2	60	8.4	0.6	7.8
4	2.4	120	16.8	1.2	15.6
8	4.8	240	33.6	2.4	31.2

*Based on f = 0.028 (turbulent flow) and Schedule 40 pipe.

Table 11-3. Flow coefficient C_V

Size/Type (inches)	Gate	Globe	Swing Check	Ball	Butterfly
1	34	10	22	90	—
2	260	35	95	500	190
4	1,150	180	410	2,500	835
8	4,850	810	1,750	9,500	3,800

Note that the values in Tables 11-1 to 11-3 are not exactly the same as they would be if only the equations presented above were used to calculate them. Actual manufacturers' published values have been used wherever possible. Also, differences in design from one manufacturer to another can cause variations in resistance values for a specific valve type. Regardless of the source of the values shown above, they are of appropriate magnitude to enable reasonable comparison between valve types.

12

Operators and Actuators

Unless otherwise specified, valves are usually supplied with and operated by a handwheel attached to the valve stem or yoke nut, or in the case of quarter-turn valves, a lever handle attached to the valve shaft. There are circumstances in which these simple devices are not adequate. These include the following:

1. When the forces required to open or close the valve are great. This occurs frequently with large valves and with valves required to operate against high fluid pressure.
2. When the time it takes to open or close the valve is longer than required.
3. When the valve must be operated from a remote location, including automatically controlled valves.

The first of these instances can be handled by the use of auxiliary gear operators that reduce the handwheel torque or lever force necessary to operate the valve. However, gear operators also increase the time it takes to open or close the valve. Making the valve operate faster, or operating it from another location, requires either a pneumatically, hydraulically, or electrically powered actuator.

GEAR OPERATORS

When it is necessary to reduce the force to manually operate a valve, it can be equipped with a spur, bevel, or worm gear operator.

142 **Operators and Actuators**

A spur gear operator is shown in Figure 12-1. Because of possible interference between the handwheel of the operator and a rising valve stem with a long travel, spur gear operators usually are used only on globe, angle, and nonreturn valves. Unlike the handwheel on these valves, which is attached to the stem, the operator drives through the yoke nut; therefore, substantial valve modification is required for spur gear operator application on these valves. Also, for the operator shown in Figure 12-1 the stem must be changed to one with a left-hand thread in order to maintain the clockwise-to-close convention of the handwheel. If the operator did not have an idler gear, this would not be necessary.

The bevel gear operator shown in Figure 12-2 is used on gate valves. The large ring gear of the operator is mounted

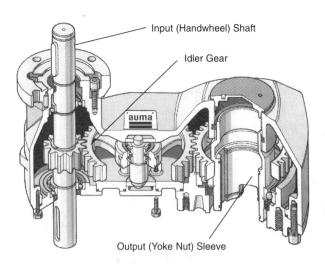

Figure 12-1. Spur gear operator

Operators and Actuators 143

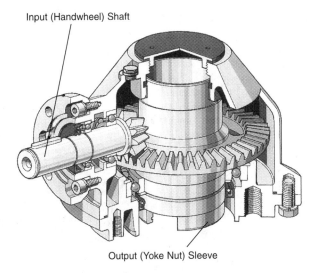

Input (Handwheel) Shaft

Output (Yoke Nut) Sleeve

Figure 12-2. Bevel gear operator

on a hollow shaft, allowing the valve stem to pass through it to reach its full travel. As the handwheel that was replaced, the operator drives through the yoke nut of the valve, thereby minimizing the valve modification required for its installation. In addition, it maintains the clockwise-to-close convention of the handwheel, making stem replacement unnecessary.

Figure 12-3 shows a typical worm gear operator. The worm gear operator is usually used on quarter-turn valves. It can be used on rising stem valves, but the high gear ratio found on this type of operator (in the range of from 20:1 to 200:1) requires an excessive number of operator handwheel turns to operate the gate or disc of the rising stem

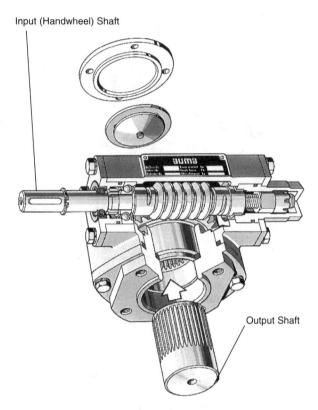

Figure 12-3. Worm gear operator

for its full travel. The worm gear operator also has the advantage of being irreversible; that is, the worm gear cannot drive the worm shaft. This is particularly important for butterfly valves in which the fluid force tends to turn the disc when it is in the partially open position. Installation of worm gear operators usually does not require extensive valve modification, only the attachment of a mounting bracket and the use of a mechanical coupling to connect the valve shaft to the output shaft of the operator. The position of the ball, butterfly, or plug of the valve is shown by an indicating arrow on the cover of the output shaft.

All three types of operators can be mounted so that the handwheel of the operator is either in line with the piping or on the side, perpendicular to the piping. Mounting of gear operators on gate, globe, angle, and non-return valves usually requires welding a pre-drilled mounting flange to the top of the valve yoke to which the operator can be bolted. Mounting operators on quarter-turn valves is accomplished by bolting a mounting bracket to the valve body by using tapped holes (such as those shown in Figures 5-2 and 7-3) provided for that purpose.

The decision as to whether a gear operator is needed—and, if so, the size of the operator required—is based on the effort required to operate the valve. The torque required at the valve handwheel or lever is compared with the torque that can be applied there by a typical valve mechanic. [One manufacturer uses 200 pound operator rim pull. This would yield a torque value of 100 × handwheel diameter (in feet) lb-ft.] If the required torque is greater, a gear operator is needed. The size of the gear operator is determined by the required torque. To simplify things for valve buyers, some manufacturers' catalogs show the sizes and pressure classes of valves that are furnished with gear operators as a standard, based on typical service pressures.

PNEUMATIC AND HYDRAULIC ACTUATORS

From an actuator applications point of view, pneumatic and hydraulic actuators can be treated as the same, only with different fluids supplying the power. Pneumatic actuators are driven by compressed gas, usually air; hydraulic actuators are powered by pressurized liquids, usually oil, although water and sometimes even the process liquid itself can be used. Because of the ready availability of compressed air in plants, pneumatic actuators are the more commonly used of the two. In the subsequent discussion, reference to pneumatic actuators should be understood to also include hydraulic actuators. There are three general styles of pneumatic actuators: linear, rotary, and linear-to-rotary conversions. Linear actuators are used on valves having translating stems. Rotary and linear-to-rotary actuators are used on valves having rotating shafts.

1. Linear Actuators. There are two different types of linear actuators, the piston and the diaphragm. The piston actuator is shown in Figure 12-4. The cylinder is fixed to the valve bonnet or yoke, and the piston rod is attached to the valve stem with the stem adapter (after the handwheel and yoke nut have been removed). Pressurizing the head end of the cylinder drives the piston, stem, and gate or disc downward, closing the valve. Pressurizing the rod end of the cylinder moves the piston, stem, and gate or disc upward, opening the valve. When fluid pressure is used to both open and close the valve, it is called a *double-acting* actuator. An alternative is the spring-return actuator. In this design an internal spring is placed between the piston and cylinder end at either the head end or rod end of the cylinder. Pneumatic pressure at the cylinder end opposite the spring moves the piston and compresses the spring. When pneumatic pressure is removed, the spring returns the piston to its original position. The spring-return actuator has the advantage of automatically closing or opening the valve if

Operators and Actuators 147

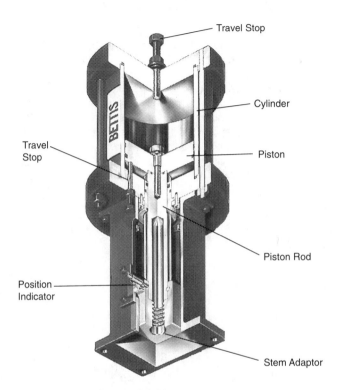

Figure 12-4. Piston actuator

operating pneumatic pressure is lost. The piston actuator in Figure 12-4 is a double-acting actuator. It has adjustable travel stops to accurately match piston travel with flow control element travel, and it has a position indicator to show stem position because the actuator housing covers the stem. The piston actuator can be used on all types of translating stem valves but is most frequently used on gate valves.

The diaphragm actuator is shown in Figure 12-5. It is similar to the piston actuator, with the non-metallic diaphragm taking the place of the piston. Diaphragm actuators can also be double-acting or spring-return actuators. Figure 12-5 shows a spring-return, "fail-closed" actuator style. This description means that if pneumatic pressure is lost, the spring will automatically close the valve. Because the diaphragm does not permit much travel, it is used only on short-stroke globe and diaphragm valves.

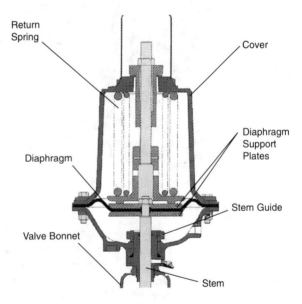

Figure 12-5. Diaphragm actuator

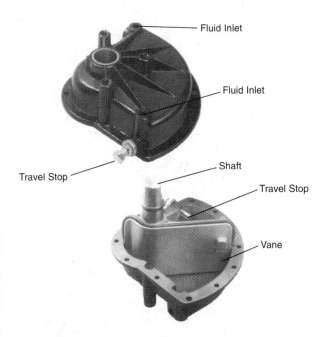

Figure 12-6. Rotary actuator

2. Rotary Actuators. The rotary actuator shown in Figure 12-6 is the rotating analog of the piston linear actuator. The quadrant-shaped housing takes the place of the cylinder, and the vane and its shaft take the place of the piston and its rod. The housing is fixed to the valve bonnet or cover, and the vane shaft is coupled to the valve shaft. The actuator in Figure 12-6 is double-acting and has travel stops to match vane travel with the flow control element travel of the valve. Because of the configu-

ration of the rotary actuator, an internal return spring cannot be readily installed. An externally mounted torsion spring that is actuated by an extension of the vane shaft must be used.

3. **Linear-to-rotary Actuators.** This class of actuators uses the piston actuator coupled with a mechanical means to convert the linear piston motion to a 90° rotating movement. Although a number of different mechanisms can be used for the conversion, the most commonly available are the rack-and-pinion and the scotch-yoke actuators. The rack-and-pinion actuator is shown in Figure 12-7, and the scotch-yoke actuator is shown in Figure 12-8.

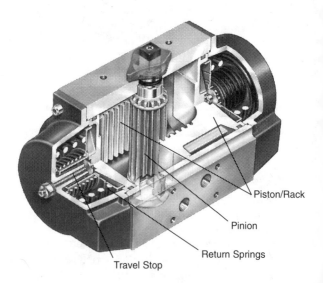

Figure 12-7. Rack-and-pinion actuator

Operators and Actuators

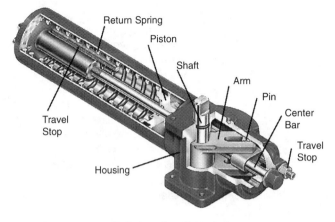

Figure 12-8. Scotch-yoke actuator

In the rack-and-pinion actuator, a gear rack machined into the extended piston takes the place of the piston rod. The rack engages a pinion gear whose shaft is coupled to the valve shaft. When the piston moves, the rack turns the pinion and valve shaft. A constant output torque at the actuator shaft is produced. As shown in Figure 12-7, dual opposed pistons usually are used to achieve balanced side-loading on the pinion and valve shafts. This actuator can be supplied in both the double-acting and spring-return designs. The actuator in Figure 12-7 has a spring-return design with travel stops to match flow control element travel.

In the scotch-yoke actuator shown in Figure 12-8, the piston slides on a fixed center bar. A pin attached to a piston extension engages a slotted arm fixed to the output shaft of the actuator. When the piston moves, the pin moves the arm, causing the actuator output shaft to turn. The pin slides in the arm slot as the arm turns. The scotch-yoke actuator produces greater torque at the beginning and end of the piston's stroke, when it is needed for seating and unseating the valve, thereby allowing a smaller cylinder to be used than that used with other mechanisms. There is no side-loading on the actuator or valve shaft, but the sliding contact causes wear of the pin and arm. This actuator can also be supplied in double-acting and spring-return designs. The actuator in Figure 12-8 has is a spring-return design with travel stops to match piston movement with flow control element movement.

All pneumatic actuators require and are usually supplied with valves to control the working fluid powering the actuator. These are small, solenoid-operated, three-way or four-way directional valves. They direct the working fluid to the appropriate actuator port (e.g., the cylinder head end) to obtain the desired valve action. The solenoids are electrically actuated from a local (at the valve) or remote control panel.

The required size of a pneumatic actuator for any specific application is determined by two factors: the pressure of the working fluid used to drive the actuator and the amount of force or torque required to operate the valve. In their catalogs, manufacturers of actuators give tables of output force or torque at different supply pressures for the different sizes of their actuators. The user must select the size of actuator that will supply the output required for the application at the working fluid supply pressure that is available.

The speed of operation of a pneumatic actuator cannot be predicted accurately. It is determined by the rate at which the working fluid enters the actuator and the ratio between the actuator output and required input. The rate at which the working fluid enters the actuator depends on the size and length of the supply line, supply pressure, and size of the solenoid valve. Consequently, faster operating times can be obtained by one or more of the following: larger supply lines, higher supply pressure, or larger solenoid valves. Using a larger actuator to produce a greater actuator-output to required-input ratio is effective only if the rate of fluid input is adequate. In the case of valves with translating stems, high-speed operation should be avoided because excessively high seating forces can result, leading to seating surface failure and jammed gate valve wedges. Regardless of the valve type, the noise and vibration (called *water hammer*) in a pipeline resulting from a valve that is closed too quickly can result in serious damage to the line.

ELECTRIC ACTUATORS

An electric valve actuator is essentially a gear operator driven by an electric motor. However, the mechanisms and controls necessary to make it practical and useful cause it to be a complex electromechanical device. A typical electric actuator is shown in Figure 12-9. Its major components, which are standard on all electric actuators, are an electric motor, a worm gear drive, travel limit switches, torque limit switches, a manual handwheel, and a position indicator.

1. Electric Motor. High starting torque, "squirrel-cage–type" alternating current motors are generally used. The high starting torque is required to overcome high starting forces in the valve, such as unseating from the closed position. The motors usually supplied are three-phase, 60-cycle, 460/230 V with thermal overload switches. Variations on the standard, such as single-

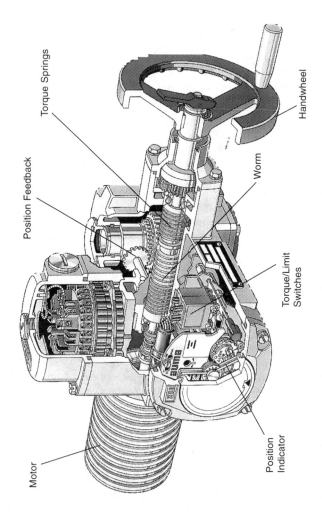

Figure 12-9. Electric actuator

phase, direct-current, 50 cycles, and other voltages, are commonly available.

Reversing the direction of actuator operation to both open and close the valve is done by reversing the direction of motor rotation. This is accomplished by using reversing motor starters that connect the motor windings properly to obtain the desired rotation.

2. Worm Gear Drive. A worm gear drive is used rather than the spur gear or bevel gear because its high mechanical advantage permits the use of low-horsepower motors. The large number of turns required to fully stroke a large valve is not the disadvantage it was in the gear operator. In fact, the worm gear drive permits the electric motor to operate at its normal high speeds. Also, the inherent irreversibility of the worm gear drive makes brakes or other positive stops to keep the valve flow control element in position unnecessary.

3. Travel Limit Switches. The cams or detents that trip travel limit switches are geared from the actuator output so that at the end of flow control element travel the switch is tripped, interrupting the control circuit and de-energizing the electric motor. A separate switch is required for each end of the travel. The switches are adjustable and are set when the actuator is mounted on the valve. The switches can also be used to energize lights or other means of valve position indication on the actuator or at a remote location.

The travel limit switch is used as the primary actuator control when the actuator is installed on a valve, which when operated manually is operated to an indicated position (i.e., quarter-turn valves).

4. Torque Limit Switches. The worm of the worm gear drive slides on its shaft between two sets of calibrated disc spring (Belleville washer) stacks. The reaction

from the worm gear causes the worm to move axially in proportion to the output torque. When the axial movement reaches a preset value, a geared cam or detent causes a switch to trip, interrupting the control circuit and de-energizing the electric motor. These switches are set at the mounting of the actuator on the valve to match the torques encountered at the full-open and full-closed positions of the valve. Because the torques encountered at intermediate positions are lower than they are at the end positions, these switches will only trip at the end positions unless an obstruction in the valve prevents full travel.

Torque limits are used as the primary actuator control when the actuator is installed on valves, which when operated manually are judged open or closed by operator "feel" (i.e., gate, global, and diaphragm valves, also known as translating stem valves).

5. Manual Handwheel. A manual mode lever simultaneously engages the handwheel drive and disengages the electric motor drive through a clutch system. Starting the electric motor automatically restores the motor drive and disengages the handwheel so that it does not turn during motor operation. Operation in manual mode is necessary during the mounting of the actuator and the setting of the limit switches. The manual handwheel also enables operation of the valve when electric power is lost.

6. Position Indication. The same gearing that drives cams or detents of the travel limit switch also drives an externally visible pointer or dial that shows the valve to be open, closed, or in an intermediate position.

More complete actuator models also include a package consisting of an integral reversing starter, a control transformer, control wiring, and local controls. The integral reversing starter allows the actuator to be completely hooked up by merely bringing the power wiring to the actuator. The control transformer steps down one phase of the power voltage to 115 V for use with the limit switches and local controls. The local controls typically consist of two three-position selector switches: local-off-remote and open-stop-close. Selecting the local position allows the actuator to be operated at the valve. Remote operation requires control wiring to be run from the actuator to a remote control panel.

Typical optional features of actuators include the following: potentiometers to provide signals proportional to valve position for remote position indication, additional limit switches for intermediate flow control element positions, timers to slow down closing speed at the end of flow control element travel in order to reduce water hammer, and position indication lights.

Most electric actuators are furnished with the valve at the time of purchase; therefore, the selection of the correct size of actuator is usually the responsibility of the valve supplier. However, it is not unusual for valves already in service to be retrofitted with electric actuators. Selecting the correct size of electric actuator to install on a valve depends primarily on two factors: the torque required to operate the valve and the desired opening or closing time. For a translating stem valve, the magnitude of the stem thrust (*force*) must also be considered.

To assist in selection, actuator manufacturers' catalogs contain tables giving output speed, rated torque, and rated thrust for the different sizes of their actuators. Generally, rated torque and rated thrust are constant for a specific size of actuator, whereas the RPM output varies in increments, depending on the internal gear ratios of the actuator. The user selects the actuator size that will provide sufficient output torque and that has an adequate thrust rating. Then, for that size, the output speed nearest to that required is selected. The required RPM value can be calculated from the desired travel time using the following formulas (with time expressed in minutes, and travel and valve stem lead expressed in inches). For quarter-turn valves,

$$\text{RPM} = \frac{1}{4 \times \text{Time}}.$$

For translating stem valves,

$$\text{RPM} = \frac{\text{Flow control element travel}}{\text{Stem thread lead} \times \text{Time}}.$$

For gate valves, the flow control element travel is approximately equal to the valve size.

VALVE FORCES

Whether it is necessary to install a gear operator or a pneumatic, hydraulic, or electric actuator on a valve, one must know the force or torque required to operate the valve. The effort required to operate a valve is used to overcome the internal forces resisting the movement of the flow control element. These internal forces fall into three categories.

First, there is the force that is unaffected by the pressure of the fluid flowing through the valve. This force would be present even if the valve were not installed in a pipeline. It is due to friction between the stem or shaft and its packing or seals. The magnitude of this force depends on the packing or seal material and how tightly the packing gland or nut is adjusted or how tightly the seals grip the shaft.

Second, there are the forces that are also present without fluid pressure and that increase as fluid pressure increases. They result from friction at the guiding and seating surfaces of the flow control element, stem thread friction, and deformation of non-metallic seating surfaces. The magnitude of these forces when there is no fluid pressure depends on the materials and finishes of the valve parts, manufacturing tolerances, and fit at assembly.

Third, there is the force that is present only when the valve is pressurized, which is the result of the fluid acting directly on the flow control element. This force is proportional to fluid pressure and can vary with the position of the flow control element in its travel.

The relative magnitudes of the different forces vary with valve type, design, and size. In some instances, the friction forces of the packing dominate; in others, the fluid-induced forces dominate.

From the previous discussions it is clear that the accurate determination of valve operating force or torque solely by calculation is not possible. Consequently, manufacturers determine the forces and torques needed to operate their valves empirically, that is, from tests they run on their products and from actual operating experience.

Manufacturers of quarter-turn valves present this information in different ways. Some use graphs showing required torque versus fluid pressure, with curves for different valve sizes. Others use tables listing required torque for different valve sizes and fluid pressures.

Manufacturers of translating stem valves generally do not provide this kind of information in their catalogs, preferring that the customer supply application data to them so that they can determine the required actuator torque or force. However, at least one valve manufacturer and some manufacturers of electric actuators provide simplified formulas and the factors to be used in them to calculate required actuator torque. These formulas take the following form:

$$\text{Torque} = (APK_1 + K_2)K_3$$

where:
- A = port area
- P = the differential pressure across the flow control element (equal to the line pressure when the valve is closed)
- K_1 = Valve Factor, a function of valve type and flow control element design
- K_2 = Packing Factor, a function of stem diameter
- K_3 = Stem Factor, a function of stem diameter, stem thread pitch, the number of stem thread starts, the coefficient of friction between the stem and yoke nut, and stem lubrication; it converts stem thrust to torque.

Valve, Packing, and Stem Factors are determined empirically by manufacturers of valves and actuators, and these factors should only be used with the manufacturers' products. However, typical values can be given. Valve factors for gate valves range from 0.2 to 0.5; for globe valves they range from 1.1 to 1.5. Packing factors range from 1,000 lbs to 5,000 lbs, increasing with stem diameter. Stem factors range from 0.006 to 0.065, increasing with stem diameter and stem lead. Because these factors must account for a wide range of applications and conditions, they generally produce actuator torques that are conservative, that is, greater than actually required.

Table 12-1 illustrates comparative actuator thrust and torque requirements for some of the different types of valves: seating and unseating thrusts and torques for gate, globe, butterfly, and ball valves. All four valves are 10-inch, Class 150, carbon steel valves carrying 200 psi clean water at ambient temperature. The torque values for the ball and butterfly valves have been taken directly from manufacturers' catalogs graphs and tables; the thrust and torque values for the gate and globe valves have been computed using actual valve dimensional data (e.g., stem diameter and port diameter) and appropriate factor values.

Table 12-1. Comparative thrusts and torques for gate, globe, butterfly, and ball valves

	Stem Thrust (lbs)	Stem/Shaft Torque (lbs-ft)
Gate*	5,425	73
Globe**	16,605	293
Butterfly[†]	N/A	210
Ball[‡]	N/A	700

*Flexible wedge, 1 3/8-inch stem.
**9.6-inch port, 1 1/2-inch stem.
[†]High-performance, TFE seat ring.
[‡]Full port, trunnion ball, TFE seat rings.

13

Control Valves and Pressure Relief Valves

The valves discussed in Chapters 2–8 have been classified by type, which depends on their flow control element. There are also two groups of valves classified by application or service, which depends on the form of fluid control they perform. These are control valves and pressure relief valves.

CONTROL VALVES

The designation *control valve* may be confusing because all valves perform a fluid control function. In the valve industry, however, the term has a specific meaning. As defined by the Instrument Society of America (ISA) S51.1, a control valve is "a final controlling element, through which a fluid passes, which adjusts the size of flow passage as directed by a signal from a controller to modify the rate of flow of the fluid."* Two things are immediately evident from this definition. First, the control valve is a component of an automatic control system. Second, the form of control performed by a control valve is regulating

---

*Reprinted by permission. Copyright 1979 (R1993). Instrument Society of America. From ISA-S51.1-1979 (R1993).

flow. Control valves are used whenever it is necessary to continuously and accurately regulate the flow rate of a fluid. This situation occurs frequently in process industries.

The type of automatic control system in which control valves are found is commonly referred to as a *regulator system*. A regulator system is designed to adjust the system output to a desired constant value and maintain it against variations caused by external disturbances. The desired constant value may be changed infrequently and over a relatively small range. A common example of a regulator system is the thermostatic control of temperature in a home or an office.

A simplified block diagram of a regulator system containing a valve is shown in Figure 13-1. The four components of the system are the controller, the valve actuator (which may be hydraulic, pneumatic, or electric), the control valve, and a flow rate measuring instrument. In this system the desired fluid flow rate is entered into the controller. The actual measured fluid flow rate is also "fed back" to the controller. If the desired and measured flow rates are different, an error signal is generated and sent to the valve actuator. In response to an error signal, the actuator turns the control valve shaft or moves the control valve stem. The valve flow control element moves, increasing or decreasing the fluid flow rate to match the desired flow rate. The flow rate measuring instrument measures the fluid flow rate downstream from the control valve and sends the value to the controller, completing the loop.

Without getting into the abstruse concepts and complex mathematics needed to understand closed-loop regulator systems, suffice it to say that for such a system to be accurate, precise, and stable, the control valves used in it must have well-defined and repeatable flow characteris-

Control Valves and Pressure Relief Valves

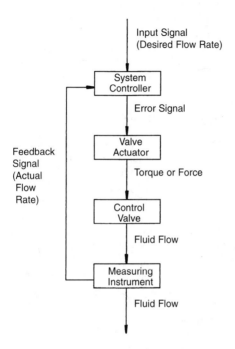

Figure 13-1. Block diagram of a regulator

tics. That is, the relationship between the position of the flow control element of a valve and the fluid flow rate through the valve must be known to a high degree of accuracy, and it must be the same every time the valve is operated. This means that the conventional valve types used for throttling services generally are not suitable for use as control valves. In reality, the typical control valve is a much modified version of one of the standard types. A detailed discussion of these valves, which have been highly engineered, is beyond the scope of this primer. However, an excellent treatment of control valve types, flow characteristics, and selection criteria can be found in *Control Valve Handbook,* Second Edition, published by Fisher Controls, Marshalltown, IA.

PRESSURE RELIEF VALVES

If it were necessary to classify pressure relief valves by type, then, because of their construction, they would be considered as highly sophisticated, spring-loaded, lift check valves. In service, however, the form of control exercised by pressure relief valves is the limiting of fluid pressure and not the prevention of flow reversal. These valves are mounted on equipment that contains fluid pressure (e.g., boilers, pressure vessels, and air receivers). They automatically open to relieve overpressure by the discharge of fluid, thereby preventing violent equipment failure, subsequent damage to adjacent equipment, and possible injury and loss of life. Pressure relief valves close and prevent the further discharge of fluid when pressures in the equipment have been restored to normal.

There are three varieties of pressure relief valves: safety, relief, and safety relief valves. They are similar in design and operation but have different applications. The safety valve is used with compressible fluids, that is, gases and vapors. The relief valve is used with incompressible fluids, that is, liquids. The safety relief valve can be used with compressible fluids and can also perform satisfactorily with liquids.

A typical safety valve is shown in Figure 13-2. During operation, fluid pressure acts on the bottom of the disc via the inlet on the bottom of the valve. Under normal conditions, the fluid force on the bottom of the disc is more than bal-

Control Valves and Pressure Relief Valves 167

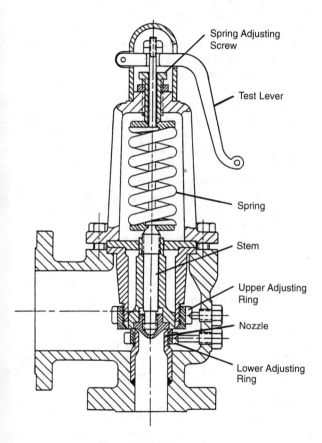

Figure 13-2. Safety valve

anced by the force of the pre-compressed precision-wound spring acting through the stem to the top of the disc. In instances in which the pressure control of the equipment on which the valve is mounted fails, the pressure increases until the fluid force on the bottom of the disc equals the spring force. At that pressure (known as the *set pressure*), the disc begins to lift. The compressible fluid then rapidly expands into what is commonly called a *huddling chamber* and acts on a substantially larger disc area. The resulting increased force greatly overbalances the spring force and fully opens the valve. All this occurs almost instantaneously, usually with a loud noise, or "pop." (Safety valves are frequently called *pop safety valves*.) With the valve open, the fluid discharges through the outlet into the atmosphere or discharge piping until the fluid pressure in the equipment is reduced to a level below the pressure at which the valve began to open. The spring force is then adequate to overcome the fluid force on the disc and close the valve.

Some of the key construction features of the safety valve are the nozzle, test lever, spring adjusting screw, and upper and lower adjusting rings. The area of the opening at the seating surface of the nozzle (called the *orifice area*) determines the relieving capacity of the valve. Pressure relief valves are manufactured with standard sizes of orifice area, designated by letter. The manual test lever provides a means for verifying whether the moving parts critical to operation are free to move. The spring adjusting screw is used to change the spring precompression and, hence, the set pressure. The pressure is set by the valve manufacturer at the factory, and the set pressure is marked on the valve nameplate. The upper and lower adjusting rings are used to adjust the configuration of the huddling chamber, regulate the amount of pressure increase over the set pressure when the valve is relieving at its full capacity (called *overpressure*), and regulate the difference between the set pressure and the reseating pressure (called *blowdown*). Inadequate blowdown inhibits firm reseating and causes the valve to "simmer" and leak. These rings usually are not adjusted unless the set pressure is changed.

A typical relief valve is shown in Figure 13-3. Its operation is similar to that of the safety valve. As in the safety valve, when the disc lifts off the seat, a larger disc area is exposed to the fluid, giving increased opening force; how-

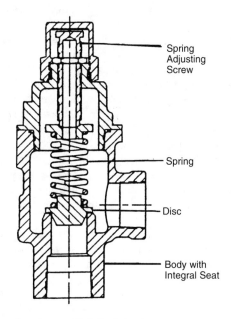

Figure 13-3. Relief valve

ever, rapid fluid expansion and subsequent full opening do not occur. The amount of opening of the relief valve is directly proportional to the increase in pressure over the opening pressure.

Construction of the relief valve is simpler than the safety valve. It does not have a huddling chamber or adjusting rings, and separate nozzles and test levers are uncommon.

Operation of the safety relief valve combines the safety valve and the relief valve. When used with compressible fluids, it pops to the fully open position, as does the safety valve. When used with liquids, its opening is proportional to the relieving fluid pressure, as in the relief valve. Its construction is similar to the safety valve. It has a huddling chamber to achieve "pop" opening when used on compressible fluids but typically has only one adjusting ring. A test lever is provided for those services, such as steam, in which it is required by applicable codes.

Although they can be used with any compressible fluid, safety valves are primarily used with steam on power boilers, superheaters, and heating boilers. They are also used on organic fluid vaporizers. Safety relief valves are typically used on all forms of pressure vessels, heat exchangers, compressed-air receivers, boiler feed-water heaters, and processing equipment. Relief valves are used at centrifugal and reciprocating pump outlets and on liquid pipelines and hot water heaters.

The design, construction, and installation of safety valves and safety relief valves used with boilers and pressure relief valves used with unfired pressure vessels are governed by the ASME in its Boiler and Pressure Vessel Code. Section I of the Code covering power boilers requires the set pressure of safety valves and safety relief valves to be at or below the maximum allowable working pressure (MAWP) of the boiler, to be fully open at no more than 3% overpressure, and to discharge all the steam from a boiler at no more than 6% over the set pressure or the MAWP. Tests to certify the capacity of the valve must be conducted at no more than 3% over the set pressure, and the blowdown cannot exceed 4% or equal less than 2% of the set pressure. Valves meeting the requirements of Section I of the Code have the ASME "V" Code symbol on their nameplate.

Section VIII of the Code requires unfired pressure vessels to be protected with a device (it permits nonclosing devices in addition to valves) that is set to operate at a pressure no higher than the MAWP and that prevents the pressure from rising more than 10% above the MAWP. Its capacity is required to be such that it can discharge the maximum quantity of fluid that can be generated without permitting a pressure increase of more than 16% of the MAWP when discharging. Valves must have their relieving capacity measured and certified at 10% over the set pressure. There are no blowdown requirements. Valves meeting the requirements of Section VIII of the Code have the ASME "UV" Code symbol on their nameplate.

In non-Code applications, it is up to the equipment designer to establish the set pressure and overpressure. Because the function of the pressure relief valve is to prevent failure of the equipment, prudence dictates that the set pressure be no higher than the MAWP of the equipment. Within this constraint, the set pressure should be high enough above the operating pressure of the equipment to prevent any unintended operation of the valve. It is generally recommended that the set pressure of safety relief valves be a minimum of 10% above the equipment operating pressure, and of relief valves a minimum of 20% above the operating pressure. Safety relief valves typically are allowed an overpressure of 10% when used on compressible fluids, and 25% with liquids. Relief valves are usually allowed an overpressure of 25%.

Determining the appropriate pressure relief valve for a specific application is a three-step process. First, the required valve type should be selected. Selection depends on the nature of the fluid to be relieved and applicable code requirements. Second, the required standard orifice size must be determined, based on the flow rate to be relieved. Manufacturers' catalogs provide equations for vapors, gases, steam, air, and liquids that can be used to calculate the minimum orifice area required for the specific conditions of relieving pressure, temperature, and so on. The standard size for this area is then selected. Alternatively, catalogs also usually provide standard orifice capacity tables for steam, air, and water from which the selection of the required orifice can be made directly. Given the required standard orifice size, a range of valve sizes and pressure classes are available. Third, the final selection is based on the same criteria that would apply to any valve, that is, fluid inlet temperature, pressure, and flow rate. Note that pressure relief valves usually have outlet sizes larger than inlet sizes, reflecting the increased volume of compressible fluids, to keep back-pressure to a minimum. Also, for flanged end valves, outlet flanges are—for obvious reasons—in a lower pressure class than are inlet flanges.

14

Selection

There are countless specific valve applications and literally hundreds of combinations of valve types, designs, materials, and so on, to choose from to fill these applications. Consequently, it is not possible to have a simple set of rules that quickly and easily identify the "correct" valve to use in any particular application.

VALVE SELECTION FACTORS

To select a valve for a specific application many factors must be considered. The most obvious ones are listed below and their effect on valve selection discussed. They have been placed into four groups: process fluid factors, process design factors, piping design factors, and economic factors.

1. Process Fluid Factors. The physical and chemical properties, temperature, and pressure of the process fluid are major determinants of the appropriate valve materials. They also establish the required pressure-temperature rating of the valve shell and can help determine valve type.

    - Whether the fluid is a liquid, gas, or vapor, and its physical and chemical properties have a direct effect on the selection of material and valve type and design. If the fluid is corrosive, corrosion-resistant shell and trim materials, such as 316SS or Alloy 20, are required. Highly erosive fluids, such as high-temperature steam, preclude the use of nonmetallic materials and may require the use of hard trim materials, such as Stellite®. Slurries require a type of valve, such as a plug or diaphragm valve, that has a smooth flow path to prevent build-up of solids. In turn, materials can limit valve type. Fluids that cannot be contaminated may require the use of a fully lined valve, such as the lined butterfly and diaphragm valves. Many more examples of the effects of fluid properties on material and type could be given, but these should be sufficient to show their importance in valve selection.

- Fluid temperature directly affects material selection. Low-temperature fluids permit the use of most materials, both metal and non-metallic. As fluid temperature increases, non-metallic materials become restricted and may be eliminated altogether, and the number of choices of metals that can be used becomes smaller. Metals such as bronze and cast iron may be eliminated.

- Consideration of fluid pressure together with fluid temperature establishes the required shell material pressure-temperature rating. This affects shell material selection and valve class. High fluid pressure alone can eliminate certain valve types such as the diaphragm valve, which is not made in high-pressure classes.

2. Process Design Factors. Process design factors—service, fluid flow rate, and frequency and speed of valve

Selection

operation—affect the selection of valve type and size and can determine whether a valve actuator is needed.

- The service in which the valve is to be used is a major determinant of valve type. In Chapters 2–8 and 13 the recommended services for each of the different types have been discussed. A summary of these recommendations is as follows:

Service	Appropriate Valve Types
Stop (on-off)	Gate, Ball, Plug, Diaphragm, Butterfly
Throttling	Globe, Butterfly, Diaphragm (weir)
Anti-backflow	Check, Non-return
Direction change	Multi-port Plug, Multi-port Ball
Pressure relief	Safety, Relief, Safety Relief

The valves used for anti-backflow, direction change, and pressure relief are well defined and are not used for other services. The types of valves that can be used for stop and throttling services are not so clear-cut. As the summary shows, only butterfly and diaphragm valves are recommended for both. Under certain conditions gate and ball valves can also be used for throttling, and the globe valve can also be used as a stop valve. When selecting a valve for stop or throttling service, the types listed in the summary for those services would be the first choice, and other types would be

chosen only if the conditions of the specific application permit.
- Fluid flow rate (e.g., gallons per minute [gpm] and cubic feet per minute [ft^3/min]) is the primary determinant of valve size. Reasonable ranges of fluid pipeline velocity for the fluids used in industrial and power piping applications have been established and are used in piping design. For a given flow rate, fluid velocity and pipeline size are inversely related, so the reasonable range of fluid velocities for the process fluid flow rate establishes a comparable range of pipe and valve sizes.
- Frequency and speed of operation influence valve type. If the valve is to be operated frequently, or the operation time of the valve is limited, or both, then the quarter-turn-type valve is preferred, especially for stopping service. Alternatively, a translating stem valve fitted with a fast-acting power actuator may be used if a quarter-turn valve is not suitable.

3. Piping Design Factors. Piping design factors are size, flow resistance, required operating force, and point of operation. All these factors affect valve type and design.

- When considering the range of sizes established by the fluid flow rate, valve type and design may be affected because not all types and designs are made in all sizes. Butterfly valves are not made in small bore sizes, globe valve and diaphragm valves are not made in very large sizes, and so on. Also, some valve designs and types, such as the butterfly and wafer check valves, take up less space in the pipeline than do others.

Selection

- Flow resistance affects the selection of types and designs of normally open stop and check valves. Those types and designs that have the lowest flow resistance would be selected.
- Some valve types and designs require greater force to operate than do others. Low operator force is preferred. If the valve requires high force, an operator or actuator may be required.
- If the valve is to be operated remotely, a power actuator is necessary, regardless of valve type. In this case, quarter-turn valves are generally easier to actuate and thus have an advantage over translating stem valves.

4. **Economic Factors.** The economic factors of cost and availability can affect valve types, designs, and materials. It is not possible to make firm statements as to which types and designs are less expensive or more available than others because both factors vary with time and location and are greatly affected by use. If a particular type and design enjoys wide use, manufacturers produce it in quantity, reducing their unit cost and selling price, and distributors stock it, making it readily available. An example of this is the conventional gate valve that is relatively inexpensive and widely available. Another example is the quarter-turn valve. Before the development of high-performance non-metallic materials, ball and butterfly valves were not very common. Today, their use rivals that of other valve types; therefore, cost is decreasing and availability increasing. Some specific valve types, designs, and materials find favor with certain industries and are readily available in areas in which those industries are located, but not in others. An example of this is the knife gate valve, which is

widely used in the pulp and paper making industry. Cost and availability affect material selection, sometimes adversely. Materials such as Monel and Hastelloy are expensive and not readily obtainable; therefore, a valve made from a material less suitable for the application may be selected to save money or meet the schedule.

Many of the factors listed previously are interdependent. Those in the last two groups are particularly related. For example, valve weight and cost are a function of valve size, which, in turn, is directly affected by the process fluid flow rate. There is flexibility among the factors in the third and fourth groups, and compromises and trade-offs are often made to reach a viable valve selection.

VALVE SELECTION PROCESS

Valve selection takes place in two situations. The first is in conjunction with the design of a new piping system. The new system can be an addition to an existing processing unit or may be one of many in a new unit or plant. The second situation is the replacement of an existing valve. Selecting a new valve for an existing system rather than replacing it in kind usually occurs when the valve being used has failed because it was improperly selected. In both cases the selection process is similar. The major difference between the two is that in an existing piping system one selection factor—size—has already been established.

The selection process begins with the factors in the first two groups, process fluid and process design. These comprise the fixed input in valve selection. In new piping systems, they are specified on a Process Flow Diagram or a Piping and Instrumentation Drawing. In existing piping systems, they are the actual process conditions. In selecting the valve, it must be remembered that the valve must be capable of meeting the temperature and pressure requirements of the fluid, the valve materials and design must be compatible with the properties of the fluid, and the valve must be a type suitable for the service. In the process of meeting the specific conditions for these factors, the valve type, design, materials, and so on, may be completely defined. In other cases, the number of possible options will be greatly reduced.

At this point in the selection process it is necessary to choose the best valve from the possible options. The third group of factors are those that must be considered and established by the piping designer in the process of the overall piping design. One of the most important aspects in the design of piping systems is establishing the pipe size. Given the reasonable range of fluid velocities, the piping designer must determine the optimum size by considering many factors. If velocities at the high end of the range are used, the piping is smaller, lighter, less expensive, and easier to install. However, small pipelines have greater resistance to flow—which causes more erosion in the pipe, fittings, and valves—and require higher horsepower pumps to drive the fluid through the system and maintain the required pressures. Thus the final pipe size is a compromise of many factors. Once the pipe size is established, however, so too is the valve size. With very few exceptions, valves should be the same size as the pipelines into which they are installed. Given its size, a valve type and design must be selected—within the restrictions established by the process conditions—that is light, has low resistance to flow, is easy to operate, and meets budget restrictions and schedule needs. The weight of the valve affects the design of the pipe supports. A light valve may be supported completely by the adjacent piping; whereas a heavy one may require its own support. If the force to operate the valve is such that an operator or actuator is required, both weight and cost are added.

As part of the selection process, preliminary cost and delivery information for the possible valve options must be obtained and incorporated into the final decision. It does no good to select the best valve for the application if its cost is over that budgeted or if the valve cannot be delivered on time. If the selected valve is not made for stock or is out of stock, then it must be custom-made by special order, which is expensive and takes time. When delivery is paramount, it may be possible to obtain a valve that is close to what is required and have it modified. Although this approach is not feasible for characteristics such as valve type, design, and body material, it can be used to change trim materials and some valve ends. If the valve is to be actuated, time must be allowed for actuator mounting.

15

Maintenance and Repair

Valves are dynamic devices that exist in a dynamic environment. They have moving parts that wear and packing and seals that also age and lose their effectiveness. They are subject to the abrasive and erosive effects of the fluids they control and the fluctuations in pressure and temperature of those fluids. Valves also experience fluctuations in environmental temperature and vibration from the pipeline in which they are located. Consequently, it is not surprising that valves need to be maintained, repaired, and sometimes replaced. Once installed, valves should be periodically inspected to determine the need for maintenance or repair. The amount of maintenance that can be performed on most valves is limited; however, repair possibilities are extensive and are limited only by economic considerations.

VALVE MAINTENANCE

Aside from lubricated plug valves, maintenance of valves is performed on an as-needed basis, and consists mainly of correcting external fluid leakage at the stem or shaft of gate, globe, ball, and butterfly valves. In most cases, stem or shaft leakage can be stopped by tightening the packing nut or gland flange nuts, compressing the packing, and forcing it tighter against the stem or shaft. In addition, although leakage of the static seals at valve body–bonnet or body–cover joints is unusual, these joints can also be tightened. On bolted-bonnet valves used in high-temperature service, bolt torque should be checked.

Lubricated plug valves need injection of sealant periodically. Sealant is displaced over time owing to fluid pressure and contact, which tends to "wash" it out. Manufacturers of lubricated plug valves recommend the frequency of seal and injection of their valves based on the frequency of valve operation—daily, weekly, monthly, or annually. Even valves that are not operated should have injection of sealant at least annually to protect the plug and body seat from corrosion. When injecting sealant, the plug should be cycled from full open to full closed to help in distributing the sealant over the seating surfaces.

VALVE REPAIR

Anything beyond the few maintenance operations mentioned previously requires at least partial disassembly of the valve and is properly termed a *repair*. Valve repairs are required for the following conditions.

1. When external leakage at the stem, shaft, or body joints cannot be stopped by tightening packing nuts, gland flanges, or body bolting. These condition are due to the failure of packing, seals, or gaskets.
2. When closing the valve does not stop fluid flow. This usually is due to the wear, corrosion, or erosion of seating surfaces. With check valves, it can be caused by the sticking and binding of the flow control element of the valve or the failure of springs used to actuate or assist flow control elements.
3. When opening the valve does not allow flow to start. This is caused by a mechanical failure of the connec-

tion between the valve stem or shaft and the flow control element.
4. When there is fluid leakage through the valve shell. This can occur in valves with cast shell components when erosion or corrosion of the inner surface of the casting uncovers porosity, which extends through the shell to the outer surface of the casting.

Some valve repairs can be done with the valve in the pipeline. The decision whether to make a valve repair in line or to remove and send it out for repair to a shop depends on the nature and urgency of the repair and on the ease of removal. In general, repair in a shop is preferred over in line repair. However, if repair **can** be done in line and must be done promptly, or if the valve is large, difficult to handle, in an awkward location, or welded into the line, in line repair may be appropriate.

The extent of in line valve repairs is limited by the type and design of the valve and whether the line has been depressurized and drained. The only in line repair that can be done while the line is still under pressure is the replacement of stem packing on gate and globe valves having backseats. After tightly backseating the stem, the valve packing nut is unscrewed, or the gland flange nuts removed and the gland flange and gland lifted (see Figures 2-2 and 2-7). The old packing can then be removed using mechanical packing pullers or high-pressure water sprays that blow out the packing. When using mechanical pullers, care must be taken not to scratch the stem or wall of the stuffing box. Scratches in these areas produce potential leak paths.

Although the old packing usually is either a continuous length of material or solid rings, the new packing must be in the form of split rings so it can be fitted around the stem.

The splits should be staggered around the stem to prevent a direct leak path. After installing the new packing the packing nut or gland flange nuts are re-attached and tightened. The stem can then be carefully unseated from the backseat, and the packing nut or gland flange nuts tightened further to stop any leakage.

Repacking a valve while it is under pressure is not a universally accepted procedure. Some manufacturers recommend against it, stating that the backseat should be used only as a temporary measure to minimize leakage until repacking can be done in a depressurized condition. Also, some valve mechanics refuse to do repacking, particularly on valves carrying steam or other high-pressure or high-temperature fluids. The decision to repack a valve under pressure depends ultimately on confidence in the backseat.

If the pipeline and valve have been drained, the possibilities for in line repair are greatly increased. The valve bonnet or cover can be removed, allowing removal of the flow control element and exposing the body seating surfaces. For certain types and designs of valves this is not possible. Titling-disc and wafer check valves, the butterfly valve, and split body and end entry ball valves cannot be disassembled in line. For valves that can be partially disassembled, the repair operations described below are feasible.

Because the stem or shaft can be removed from the bonnet or cover, replacing stem packing is greatly facilitated. In addition, the new packing can be continuous material or solid rings. It is considered good practice to replace packing, seals, and gaskets whenever a valve is disassembled for any repair.

Seat rings can be replaced or refinished. For top-entry and three-piece ball valves, the removal and replacement of seat rings is easily accomplished. Separate seat rings on gate, globe, and check valves can be replaced if they are screwed in; however, special spanner wrenches may be required. If seat rings are tack-welded to the body, the tack weld must be cut away. Welded-in seat rings or integral seating surfaces can be refinished with the valve in place. Doing so requires special machinery that clamps onto the valve and "pilots in" the seat ring bore or some other surface concentric to it. Because seat rings on globe and lift check valves are directly below and parallel to the body opening, good results are relatively easy to obtain. The seating surfaces on gate and swing check valves are off-set from and at an angle to the body opening; therefore, good results are more difficult to obtain. Companies specializing in in line valve repair usually are brought in to do repair operations that require special machinery and know-how. After replacing or refinishing the body seats, their fit with the flow control element of the valve must be checked. It may be necessary to do further grinding or lapping of the seat rings or flow control element to get a good, leak-tight fit.

Because it has been removed from the body, replacing or refinishing a flow control element, stem, or shaft is easily accomplished. If the flow control element is a gate valve wedge, globe valve disc, or check valve disc or piston, its fit with the seat ring(s) must be checked, even if it is a new replacement part. Ball valve balls cannot be refinished easily, and therefore they usually are replaced. However, this is seldom necessary because the soft seat rings in these valves experience almost all the wear. Diaphragm valve diaphragms are easily replaced, and the fit with their seating surfaces is not a concern.

After an in line repair the valve should be pressure tested to check the integrity of the body–bonnet or body–cover joint and seat tightness. Because this would involve all or at least part of the pipeline, it may not be possible to test at the pressures that would be used to test the valve alone (normally 1½ times the CWP of the valve). The recommended practice is to test at 1½ times the design pressure of the pipeline. Checking for seat leakage is also difficult after an in line repair. Because the seating surfaces are not visible, seat leakage must be checked at a point downstream of the valve.

SHOP REPAIR

Valve repair in a shop has distinct advantages over in line repair. The type of repair is not limited to those that are done in line. All types and designs of valves can be repaired, and repairs such as welding of leaking body castings can be done more easily. Repairs are limited only by the capability of the shop. Because there is much better control over the work the quality is better, and the repaired valve can be more effectively tested for shell and seat tightness. Furthermore, if the valve can be replaced with a spare one, the downtime of the piping system may be less than with an in line repair, and the repaired valve can then become a spare.

There are circumstance other than the immediate need for a repair in which valve repair shops are used extensively. Power plants, oil refineries, and large chemical plants have periodic shut-downs or turnarounds during which all or some of the boilers or processing units are taken out of service for scheduled maintenance and deferred repairs. It is common practice during these stoppages to remove all the large bore valves and send them to valve repair shops for disassembly, inspection, and if necessary repair. For economic reasons, small bore piping systems, including their valves, are usually scrapped and replaced rather than repaired, unless the system is fabricated of expensive or hard to obtain materials.

For the repair of pressure relief valves, selection of a valve repair shop may be restricted. Only repair shops holding National Board of Boiler and Pressure Vessel Inspectors "VR" certification are authorized to make repairs to ASME "V" and "UV" stamped valves.

Work done in a valve repair shop should be controlled by a valve repair specification, which defines the repairs permitted and the quality of work required. Most large power companies, oil companies, chemical companies, and so on, have valve repair specifications. Although there are variations in the specific requirements in these specifications, they all address the following topics.

1. Disassembly and cleaning. The match-marking of parts to ensure proper reassembly and the methods permitted for cleaning without damaging them are addressed.
2. Inspection. The visual and dimensional standards to be used to check cleaned parts are defined. (See Appendix 1 for a listing of applicable standards.)
3. Evaluation of inspection results. The criteria for the rejection of inspected parts and the valve as a whole are stipulated. A limit usually is set on the allowable cost of repair of a valve. A percentage of the cost of a new valve usually is established. If the cost of repair exceeds the percentage, the valve is scrapped.
4. Permitted repairs. The repairs allowed for specific valve parts are defined. They include requirements for the welding of parts, refinishing or weld overlaying of seating surfaces, machining of stems and shafts, machining of flange faces, and so on.

5. Reassembly. The need to replace packing gaskets, seals, and bolting is stated, and the specific requirements for these parts are defined. Dimensional requirements for re-assembled valves are also specified.
6. Testing. The need to pressure test each repaired valve is stated, and specific testing requirements are defined. Reference usually is made to an established testing standard such as API 598. Such standards define the different tests required, test media, pressure duration, and allowable leakage.
7. Preparation for shipment. Painting and tagging requirements and the records necessary to document the work performed are established.

Using valve repair specifications produces repaired valves that—although not equal to new ones in all respects—can be re-installed in the services from which they were removed and give many additional years of satisfactory service. At the same time, using repaired valves reduces the expense of keeping power plants, oil refineries, and chemical plants in operation.

16

Miscellaneous Topics

BYPASSES AND DRAINS

A valve bypass is a small pipeline connecting the inlet and the outlet of a valve, going completely around the flow control element. The bypass incorporates a small valve, usually a gate valve, that is normally closed. It is standard practice to attach the bypass at the side of the main valve with the stems of both valves parallel, pointing vertically upward, as shown in Figure 16-1.

Bypasses serve two purposes. First, by opening the bypass valve the upstream pipeline pressure is reduced and the downstream pipeline pressure increased, thereby reducing the decrease in pressure across the main valve, the fluid force on the flow control element, and the force required to open the main valve. Second, on steam and other high-temperature services, opening the bypass valve allows a small flow of hot fluid to warm the downstream pipeline before the main valve is opened, thereby reducing thermal shock downstream. In both instances the bypass valve is normally closed after opening the main valve.

The size of a bypass line is determined by the size of the main valve and the purpose of the bypass. Balancing of pressures requires larger lines than does pipeline warming. MSS Standard SP-45 specifies the sizes of bypass lines for standard installations.

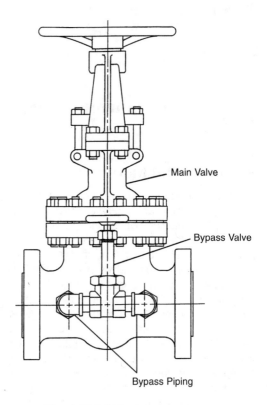

Figure 16-1. Valve bypass

A drain is a hole in the valve body, usually at the bottom, that is tapped and fitted with a pipe plug. It is used to empty fluid from a depressurized valve and those parts of the connecting pipeline that are at a higher elevation. It is also used to drain fluid from between the seats of double-seated valves in block-and-bleed operations and to relieve body cavity pressure. If draining is done frequently, a pipe nipple and a small stop valve are installed instead of a pipe plug.

To enable bypass and drain installation, valve body castings made of steel and cast iron usually are designed with bosses at appropriate locations to provide sufficient material to have the required thread length for a threaded connection, or to allow for the depth of a socket-weld socket and still have the required valve shell wall thickness.

EXTENDED BODY VALVES

Generally, valves are symmetrical end-to-end; that is, the flow control element is midway between the ends. Extended body valves are the notable exception to this generalization. They are small, forged steel, gate and globe valves in which one end is substantially extended. An extended body valve is shown in Figure 16-2. These valves are used in applications in which venting, draining, and sampling connections are made directly on vessels, pipe headers, and so on. They take the place of a nipple and standard valve assembly, eliminating a weld or threaded connection, and they are stronger.

The valve shown in Figure 16-2 has a welded bonnet, but extended body valves are also made with bolted bonnets. There are a number of options available for ends. The valve in Figure 16-2 has an integrally reinforced male couplet on the extended, or male, end. The male couplet end is butt-welded directly to the wall of the header or vessel to which the valve is being attached. The small ring extended from the end is inserted into a hole in the wall and acts as a backing ring for the attaching weld. Other extended end options are the male pipe thread end and the male socket-weld end (equivalent to pipe). These are also

Miscellaneous Topcs

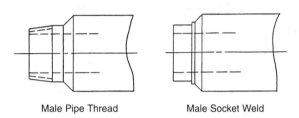

Male Pipe Thread   Male Socket Weld

Figure 16-2. Extended body valve

shown in Figure 16-2. The male pipe thread end is screwed into a transition piece, commonly known as a *threadolet,* that has been welded to the vessel or header wall. Similarly, the male socket-weld end is fillet-welded into a transition piece, known as a *sockolet,* that has been welded to the vessel or header wall. The short, or female, end of the extended body valve can be a threaded end or socket-weld end.

DOUBLE-SEATED VALVES

Valves with two seating surfaces, such as gate and ball valves, can trap pressurized liquids and gases in the cavity between the seats when the valve is closed. This can have undesirable results. If the temperature of this trapped fluid is increased, the pressure also increases. If the increase in pressure is great, damage or deformation of sealing materials, body–bonnet bolting, or the valve shell can result. Although the number of applications in which this situation can occur is not great, these applications should be recognized and one of the following preventive measures taken before the valve is put into service.

1. Drill a pressure equalization hole through the upstream side of the gate valve wedge or the ball valve ball. This makes the valve unidirectional.
2. Install a small pipeline between the body cavity and the upstream end of the valve. This also makes the valve unidirectional.
3. Install a drain plug through the body cavity wall of the valve, and remove the plug when the valve is closed. This is only feasible if the valve is rarely closed, and care must be taken to reinsert the plug before the valve is opened.

4. Install a pressure relief valve through the body cavity wall of the valve.

Some manufacturers of ball valves have recognized the need for cavity pressure relief and designed seat rings that move slightly away from the surface of the ball when cavity pressure exceeds line pressure, allowing cavity pressure to relieve into the valve bore.

Even if cavity pressure build-up does not require special measures, if the valve is kept in the closed position, fluid trapped in the valve body cavity remains under pressure after the piping system has been depressurized and even if the valve is removed from the piping system. This valve can be a hazard during maintenance operations. If it is removed from service in the closed, pressurized position and then is opened, or the bonnet unbolted, significant quantities of fluid will be released. If the fluid is toxic or flammable, serious injury can result. Furthermore, the act of removing the bonnet in itself can be dangerous.

SOUR SERVICE VALVES

Metallic materials in contact with sour environments (fluids containing water as a liquid and hydrogen sulfide gas with certain partial pressures and concentrations, or sour crude oil with certain hydrogen sulfide gas concentrations and pressures) are subject to a form of corrosion known as sulfide stress cracking (SSC). These sour environments occur frequently in oil production and refining. The National Association of Corrosion Engineers (NACE) Standard MR-01-75 addresses in detail the specific environments in which SSC must be considered in material selection.

Sulfide stress cracking can be controlled through the use of materials that have been properly heat treated and are of the appropriate hardness. NACE Standard MR-01-

75 specifies the heat treatments and hardness requirements for a wide range of engineering alloys, including those used in the manufacture of valves.

Most manufacturers offer valves made of materials that meet NACE Standard MR-01-75 requirements; however, as in all instances in which special requirements are invoked, these valves are not readily available in all sizes and pressure classes. In addition, because there are different heat-treatment requirements for different materials, some manufacturers do not offer all trim options on valves whose materials must meet NACE requirements. A prime example of this is the lack of availability of 13% chromium stainless steel trim. In order to meet NACE requirements this martensitic stainless steel must undergo a three-step heat-treatment process; whereas other trim materials, such as 316SS and Monel, can be used in the annealed condition or need only meet hardness requirements.

FIRE SAFE VALVES

Despite the best preventive efforts, fires do occur in oil refineries and chemical plants. In these instances, soft-seated valves, such as ball and butterfly valves, present a unique problem. During a fire temperature can rise above the maximum operating temperatures of the materials used to make seats and seals, causing them to decompose or deteriorate and sometimes producing disastrous results. If the valve was closed, it will no longer be closed. The effect of this uncontrolled opening depends on the specific application; however, it is rarely beneficial. More importantly, loss of stem and body seals produces leakage directly to the environment, and if the valve is in a service such as furnace fuel lines, LP-gas lines, and petroleum and petrochemical piping systems, the leaking fluid literally adds fuel to the fire.

In order to prevent such occurrences, designers of ball and butterfly valves have modified their designs to reduce leakage, assuming the absence of the non-metallic seats and seals. For ball valves, the following modifications are typical:

1. On floating ball designs, a secondary integral metal seat is machined into the valve body adjacent to the soft seat. The slotted stem-to-ball connection allows line pressure to move the ball downstream when in the closed position and bear against this seat for closing.
2. Metal, spiral-wound graphite-filled gaskets are used as body seals in split body and top-entry valves, rather than the non-metallic materials usually used.
3. A backseat is machined in the body or cover of the valve that mates with an integral collar on the internally inserted shaft to produce a metal-to-metal seal.
4. High-temperature graphite shaft packing is used in lieu of the non-metallic packing usually provided.

The usual modification for butterfly valves is to encase the non-metallic seat ring in a metal carrier, which becomes a secondary seating surface in the absence of the non-metallic material. Graphite shaft seals are also used instead of the usual non-metallic seals.

To verify the effectiveness of these measures, several standards, such as API 607, have been developed. API Standard 607 specifies a test that simulates typical plant fire conditions and stipulates criteria for evaluating valve performance both during and after the test. Maximum internal and external leakage rates are specified. The Standard also provides for extending the test results to other sizes and ratings of valves that have the same basic design and the same non-metallic materials. Valves that have satisfied the prescribed performance requirements can be certified by the manufacturer as "Fire Tested per API 607."

LOW FUGITIVE EMISSIONS VALVES

The term *fugitive emissions* applies to leaks from process equipment such as pumps, valves, and compressors and from the piping connections between them, as opposed to the controlled emissions from stacks and vents.

Federal, state, and local agencies have issued regulations restricting the fugitive emissions of volatile organic compounds (VOCs). VOCs can be loosely defined as carbon-containing compounds that exist as a gas at ambient pressure and temperature. Some 149 different toxic VOCs are covered by these regulations. For valves, leakage is determined by measuring the local VOC concentration at the valve surface at a possible leak source (e.g., a packing gland or a body–bonnet joint) with an organic vapor analyzer. There is no uniformity among federal, state, and local regulations as to the level of concentration that is unacceptable. However, currently the 1990 US Environmental Protection Agency's Clean Air Act defines an unacceptable leak for valves as 500 parts per million (ppm) or more of VOCs. Local and state regulations as to concentrations may be lower.

Valve manufacturers have taken several different approaches to reducing fugitive emissions, among which are the following:

1. Using specially engineered packing sets that have the capability to expand radially up to $1/8$ inch upon gland tightening to achieve improved leak-tight sealing against the stuffing box and the valve stem.

2. Redesigning stuffing boxes to achieve more uniform compression of the packing and better sealing. Design changes include shorter, narrower stuffing boxes; tighter tolerances on stem or shaft diameter and straightness, and on stuffing box diameter and concentricity; and improved finishes on stems, shafts, and stuffing boxes.
3. Using stuffing boxes with two sets of packing separated by a lantern ring and a leakoff connection for the detection of a leak in the lower packing set. The upper packing set prevents leakage to the environment until the valve can be repacked. The leakoff connection must be monitored.
4. Live-loading the packing gland by installing springs under the bolt nuts of the gland flange. These springs help maintain packing compression after cycling and aging of the packing. A live-loaded butterfly valve is shown in Figure 5-2.
5. Using stuffing boxes with two sets of packing separated by a lantern ring and equipped with an injector for the injection of sealant between the sets of packing. The injected sealant fills the leak paths at the outer diameter of the stem and the inner diameter of the stuffing box. This seal must be maintained by replenishing the sealant periodically, and the sealant must be compatible with the fluid.
6. Using a metal bellows to seal the stem enclosure. A bellows seal globe valve is shown in Figure 16-3. The stem is encased in a cylindrical bellows that has one end seal welded to the lower end of the stem and the other end seal welded to the bottom of the bonnet. The bellows compresses as the valve is opened and expands as the valve is closed. The result is a completely leak-tight assembly. Stem packing is retained as a back-up in the

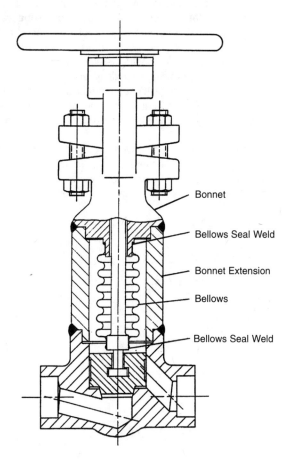

Figure 16-3. Bellows seal valve

event of bellows failure. Metal bellows can only be used with non-rotating stems. When used on globe valves the stem must be driven by the yoke nut, as in a gate valve. Bellows seal valves have a history of successful use in services in which fluid leakage to the atmosphere is unacceptable owing to toxicity, chemical aggressiveness, radioactivity, and other health hazards. Bellows seals usually are found only on small forged steel valves. They add substantially to the cost of a valve.

All of these approaches, whether used separately or in combination, have been effective in reaching acceptable concentrations of fugitive emissions, but only the last one—the bellows seal valve—achieves zero fugitive emissions.

Appendix 1

Standards

Listed below are some of the standards issued by national organizations that are applicable to the types of valves discussed in this primer. These standards establish requirements for the design, dimensions, materials, marking, inspection, and testing of valves. Most of them are directed primarily toward manufacturers; however, they are also an excellent source of information for users.

American Society of Mechanical Engineers (ASME):

- B16.10 Face-to-Face and End-to-End Dimensions of Valves
- B16.34 Valves—Flanged, Threaded, and Welding Ends
- Boiler and Pressure Vessel Code:

    Section I – Power Boilers
    Section IV – Heating Boilers
    Section VIII, Division 1 – Pressure Vessels

American Petroleum Institute (API):

- 526 Flanged Steel Safety Relief Valves
- 594 Wafer and Wafer-Lug Check Valves
- 598 Valve Inspection and Test

Appendix 1

- 599 Steel and Ductile Iron Plug Valves
- 600 Steel Gate Valves, Flanged or Butt Welding Ends
- 602 Compact Steel Gate Valves—Flanged, Threaded, Welding, and Extended Body Ends
- 603 Class 150, Cast, Corrosion-Resistant, Flanged-End Gate Valves
- 604 Ductile Iron Gate Valves—Flanged Ends
- 607 Fire Test for Soft-Seated Quarter-Turn Valves
- 608 Metal Ball Valves Flanged and Butt-Welding Ends
- 609 Butterfly Valves, Lug-Type and Wafer-Type

Manufacturers Standardization Society of the Valve and Fittings Industry (MSS):

- SP-25 Standard Marking System for Valves, Fittings, Flanges, and Unions
- SP-42 Class 150 Corrosion Resistant Gate, Globe, Angle and Check Valves with Flanged and Butt Weld Ends
- SP-45 Bypass and Drain Connections
- SP-61 Pressure Testing of Steel Valves
- SP-67 Butterfly Valves
- SP-68 High Pressure Offset Seat Butterfly Valves
- SP-70 Cast Iron Gate Valves, Flanged and Threaded Ends
- SP-71 Cast Iron Swing Check Valves, Flanged and Threaded Ends
- SP-72 Ball Valves with Flanged or Butt-Welding Ends for General Service

Appendix 1

- SP-78 Cast Iron Plug Valves, Flanged and Threaded Ends
- SP-80 Bronze Gate, Globe, Angle, and Check Valves
- SP-81 Stainless Steel, Bonnetless, Flanged, Knife Gate Valves
- SP-82 Valve Pressure Testing Methods
- SP-84 Valves—Socket Welding and Threaded Ends
- SP-85 Cast Iron Globe and Angle Valves, Flanged and Threaded Ends
- SP-88 Diaphragm Type Valves
- SP-91 Guidelines for Manual Operation of Valves
- SP-92 MSS Valve User Guide
- SP-99 Instrument Valves
- SP-101 Part-Turn Valve Actuator Attachment
- SP-102 Multi-Turn Valve Actuator Attachment
- SP-110 Ball Valves Threaded, Socket-Welding, Solder Joint, Grooved and Flared Ends

National Association of Corrosion Engineers (NACE):
MR-01-75 Sulfide Stress Cracking Resistant Metallic Materials for Oil Field Equipment

The standards marked with a square are also American National Standards Institute (ANSI) standards.

Appendix 1

Engineering standards and specifications referenced in *The Valve Primer* are available from the appropriate society or association at the following addresses:

American National Standards Institute (ANSI)
11 West 42nd Street
New York, NY 10036

American Petroleum Institute (API)
1220 L Street NW
Washington, DC 20005

American Society for Testing and Materials (ASTM)
1916 Race Street
Philadelphia, PA 19103

American Society of Mechanical Engineers (ASME)
345 East 47th Street
New York, NY 10017

American Welding Society (AWS)
550 LeJune Road NW
Miami, FL 33135

Instrument Society of America (ISA)
67 Alexander Drive
Research Triangle Park, NC 27709

Manufacturers Standardization Society of the Valve and Fittings Industry (MSS)
127 Park Street NE
Vienna, VA 22180

National Association of Corrosion Engineers (NACE)
1440 South Creek Drive
Houston, TX 77084

Appendix 2

Glossary

Acme thread A flat-topped trapezoidal shaped thread used for power transmission rather than fastening.

actuator A mechanical device that uses an energy source other than manual power to operate a valve; also called an operator.

adjusting rings The parts of a safety valve or safety relief valve used to control disc lift and blowdown.

ambient temperature The temperature of the environment surrounding a valve, normally ranging from −20°F to 100°F.

angle valve A globe valve whose ends are perpendicular to each other, rather than parallel; one end is directly below the disc.

backseat An auxiliary seat in the bonnet of a gate valve or a globe valve that provides a seal between the stem and bonnet to enable packing replacement while the valve is under pressure.

ball The spherical flow control element of a ball valve or ball check valve.

ball check valve A lift check valve whose flow control element is a solid ball.

ball valve A type of valve whose flow control element is a ball with a circular passage through it and that rotates $90°$ from open to closed.

bellows seal valve A gate or globe valve that uses a cylindrical metal bellows to hermetically seal the valve against stem leakage.

bevel gear operator A gear operator that uses a bevel gear set.

block-and-bleed An operation done with a double-seated valve having a between-seats drain. When the valve is closed (blocked) the drain is opened, allowing the fluid trapped between the seats to drain (bleed) away.

blowdown The difference between the set (opening) pressure and the disc reseating pressure of a pressure relief valve, which is usually expressed as a percentage of the set pressure.

body The part of a valve that houses the flow control element, contains seating surfaces, retains fluid pressure, and in most valve designs, has ends for attaching to connecting pipe.

bolted bonnet A body–bonnet joint design in which the bonnet is fastened to the body using studs or bolts and nuts; it uses a gasket to provide joint sealing.

bonnet The part of a gate valve, globe valve, or diaphragm valve that is fastened to the body to complete the pressure-retaining shell; it has an opening for the stem to pass through, and it usually contains a stuffing box. Depending on the valve stem design, it may also have a yoke attached.

bore The diameter of the smallest opening through a valve; also called a port.

bubble tight A phrase used to describe the tightness of valve seating surfaces. It derives from a seat tightness test in which compressed air is used for the test, with water on the downstream side of the seat so that leak-

Appendix 2

age is detected by air bubbles. It is usually used in conjunction with soft-seated valves to indicate zero leakage.

butt weld end A valve end machined to the proper dimensions to enable butt welding to connecting pipe.

butterfly valve A valve whose flow control element is a disc with an axis perpendicular to flow and that rotates 90° from open to closed.

bypass A short pipeline, containing a stop valve, that is mounted on a valve to connect the inlet and the outlet and bypass the flow control element.

C_v *See* flow coefficient.

cap The part of a check valve, ball valve, or plug valve that is fastened to the body to complete the pressure-retaining shell; on ball and plug valves the cap has an opening for the stem to pass through and may also contain a stuffing box; also called a cover.

chain wheel An assembly consisting of a sprocket wheel, chain guide, and chain that is fastened to the handwheel of a gate valve or a globe valve; it is used to operate overhead or inaccessibly mounted valves. The assembly is mounted with its stem horizontal, with the chain hanging down.

check valve A type of valve that permits fluid flow in one direction only; back-flow is prevented. There are two basic styles: the swing check valve, in which the flow control element rotates around an axis; and the lift check valve, in which the flow control element translates along the fluid path.

clapper *See* flapper.

class A system used to categorize valves according to their pressure-retaining capabilities.

cock A small plug valve. It is also called a plug cock.

cold working pressure The maximum pressure-retaining capability of a valve at ambient temperature. It is marked on the valve as CWP. It is also known as the Water-Oil-Gas (WOG) rating.

conical seat A design of a globe valve seat in which the seating surfaces of both the disc and the seat ring are cone-shaped.

control valve A valve used as a component of an automatic control system; it continuously and accurately controls the flow rate of a fluid. Generally, it is a butterfly valve, a diaphragm valve, or modified form of a globe valve or a ball valve.

controller A component of an automatic control system; it compares input and feedback signals and generates an error signal based on the difference between the two.

cover *See* cap.

cylinder actuator A linear valve actuator that uses a pneumatic or hydraulic cylinder and piston to produce linear motion.

diaphragm The flow control element of a diaphragm valve or the working element of a diaphragm actuator. A flexible membrane that separates regions having different pressures.

diaphragm actuator A linear valve actuator that consists of a housing divided into two chambers by a diaphragm, which is attached to a stem, which passes out of one chamber; pressurizing either chamber produces linear motion of the actuator stem.

diaphragm valve A type of valve that uses a diaphragm as a flow control element; the diaphragm lines one side of the fluid path and is pushed across the fluid path.

Appendix 2

disc The flow control element of a globe valve, check valve, or butterfly valve.

double-acting actuator An actuator that uses external energy to both open and close the valve.

double-disc gate A flow control element of a gate valve that has two separate, parallel seating surfaces.

double-seated valve A valve that has two separate seating surfaces in its body; the flow control element comes into contact with both seats when the valve is closed.

drain An opening in the valve body for the removal of fluid from the valve or the pipeline; it may be filled with a removable plug or with a pipe nipple and a stop valve that is normally closed.

E.M.O. Electric motor operator; *see* electric actuator.

elastomer A synthetic non-metallic material with properties similar to those of natural rubber.

electric actuator An actuator that uses an electric motor to supply external energy to the valve; also called an electric motor operator or an E.M.O.

equal percentage flow characteristic A flow characteristic in which equal increments of flow control element travel distance produce equal percentage changes in the flow rate through the valve; changes in the flow rate are small at the start of flow control element travel (when the valve begins to open) and become increasingly larger.

equivalent pipe length A measure of a valve's resistance to flow where the resistance is equal to and expressed as a length of pipe of the same size as that connected to the valve.

extended body valve A small forged-steel gate or globe valve that has a body with one end extended substantially for direct connection to pressure vessels, piping headers, and so on, without the use of a pipe nipple.

fire safe   A designation applied to a valve that has been verified by testing to adequately seal against external and internal leakage during a fire; it is usually used in conjunction with a soft-seated valve.

flanged end   A valve end that has an integral flange for bolting to a connecting pipe having a similar flange.

flapper   The flow control element of a swing check valve; also called a disc or a clapper.

flat seat   A seat for a globe valve in which the seating surfaces of the disc and seat ring are flat rings that are perpendicular to the direction of flow.

flexible seat   A seat design of a ball valve that provides controlled deformation of the non-metallic seat rings when the valve is assembled.

flexible-wedge gate   A wedge design flow control element of a gate valve made of a single piece that has a groove around its perimeter to provide flexibility and some movement of its seating surfaces; also called a flex wedge.

floating ball   A flow control element of a ball valve that is held in position in the valve body solely by the seat rings of the valve.

flow characteristic   The relationship between the fluid flow rate through a valve and the position of the flow control element of the valve.

flow coefficient   A measure of the flow capacity of a valve; designated as C_v, it is the flow rate in gallons per minute (gpm) of 60°F water that passes through the fully open valve at 1 psi (pounds per square inch) pressure differential.

flow control element   The part of a valve that obstructs and controls fluid flow; it determines the valve type and the nature of fluid control for which the valve is suited.

fugitive emissions   A name used by environmental protection agencies for the external leakage of hazardous gases from piping components (e.g., valves and pumps).

full bore   A bore of a valve that is approximately the same size as the inside diameter of the connecting pipe; also called a full port.

full port   *See* full bore.

gasket   A flat, yielding, ring-shaped part placed between mating components (e.g., body and bonnet, flanges) in pressure-containing assemblies to form a fluid seal.

gate   The flow control element of a gate valve; also called a wedge or slide.

gate valve   A type of valve in which the flow control element enters the fluid path from the side and traverses it.

gear operator   The operator of a valve that uses a gear set to reduce the force required to close the valve.

gland   A tubular part that fits over a valve stem or shaft that is used to compress packing. Also called a gland bushing.

gland flange   A part that fastens to the bonnet of a valve used to apply force to the gland. It is usually found on valves with outside-screw-and-yoke (OS&Y) stem designs.

gland nut   A nut that threads onto or into the top of the bonnet of a valve to apply force to the gland.

globe valve   A type of valve in which the flow control element moves parallel to the direction of fluid flow; its name derives from the spherical shape of its body.

grooved end   A valve end that has a circumferential groove around it to receive a clamping mechanism, which couples the valve to similarly grooved connecting pipe.

hard facing   A hard material, such as Stellite®, that is deposited on a relatively soft base metal by welding to produce surfaces that are wear and erosion resistant.

head   Fluid pressure expressed in terms of the height of a column of the fluid.

high-performance butterfly valve   A butterfly valve in which the disc seating surface is off-set from its axis of rotation, producing an uninterrupted seating surface and "camming" action when seating and unseating.

huddling chamber   A space adjacent to the seating surfaces of a safety valve or a safety relief valve that captures high-pressure fluid at disc unseating, producing higher unseating force and instantaneous full valve opening.

hydraulic actuator   A valve actuator that uses pressurized liquid as its energy source.

inherent flow characteristic   The flow characteristic of a valve if the pressure drop across the valve is held constant.

inside rising stem (ISRS)   A stem design in which male threads on the stem mate with female threads in the valve bonnet; turning the stem causes it to move out of (rise) or into the valve.

installed flow characteristic   The flow characteristic exhibited by a valve when the valve is installed in a pipeline. (The pressure drop across the valve varies.)

iron body, bronze-mounted   A phrase used to describe a cast iron valve with bronze trim.

jam seat   A ball valve seat design in which the nonmetallic seat rings are compressed (jammed) by the ball when the valve is assembled.

knife gate valve   A gate valve design that is distinguished from the standard design by the use of a simple metal plate for the gate and the absence of a bonnet; also called a slide valve.

laminar flow   Fluid flow where fluid "particles" move in definite paths parallel to the overall direction of flow; also called viscous flow and streamline flow.

lantern ring   A metal spacer ring placed in a stem packing set so that it lines up with a leakoff connection in the bonnet; it divides the packing into an upper set and a lower set.

leakoff connection   A threaded hole through a valve bonnet wall in the area of the stuffing box; it is used to collect leakage past a lower packing set or to inject lubricant into the stuffing box.

lift check valve   A type of check valve in which the flow control element moves parallel to the direction of fluid flow; the force of the fluid lifts the flow control element off its seat.

linear flow characteristic   A flow characteristic where equal increments of flow control element travel distance produce equal increments of change of flow rate through the valve. Flow rate is directly proportional to the position of the flow control element.

linear valve actuator   An actuator that produces linear motion for use with valves having translating stems, that is, gate, globe, and diaphragm valves.

linear-to-rotary actuator   A valve actuator that converts the output of a linear device to rotary motion for use with valves having rotating shafts, that is, ball valves, butterfly valves, and plug valves.

lined butterfly valve   A design of a butterfly valve that has a body liner and a disc whose axis of rotation passes through the seating surfaces of the disc and body (generally referred to as a "conventional" disc).

liner A non-metallic covering applied to the inside of the body of a valve. Liners are used mostly on butterfly valves, plug valves, and diaphragm valves for corrosion resistance.

lubricated plug valve A design of a plug valve that uses injected sealant (lubricant) to reduce the required operating force and achieve a tight seal against internal leakage.

lug body A valve body that does not have ends but has lugs (bosses), with threaded holes around its perimeter for fastening to pipe flanges.

multi-port valve A ball valve or a plug valve with more than two ends that is used for changing flow direction.

needle valve A small globe valve without a separate disc, but with a stem with an integral conical seating surface.

negative position The installation position of a multi-port valve such that fluid inlet pressure can be into a closed body opening, thereby retarding valve sealing.

non-return valve *See* stop-check valve.

non-rising stem (NRS) A stem design used on gate valves in which male threads on the stem mate with female threads in the valve gate; turning the stem causes the gate to move, but the stem does not translate (rise).

operator A mechanical device used to reduce the force required to operate a valve manually; also called an actuator.

outside-screw-and-yoke (OS&Y) A stem design in which the threaded portion of the stem is outside the pressure boundary of the valve; it must be used with a bonnet having a yoke, which holds a yoke nut to allow valve operation.

overpressure A pressure increase over the set pressure of a pressure relief valve; it is usually expressed as a percentage of set pressure.

packing Deformable non-metallic material formed into rings used for sealing around the stems and shafts of valves.

packing chamber *See* stuffing box.

plastic A synthetic non-metallic material possessing structural rigidity.

plate The flow control element of a wafer check valve.

plug The flow control element of a plug valve.

plug cock A small plug valve. Plug cocks have no provision for reducing friction between the plug and body.

plug valve A type of valve using a cylindrical or conical flow control element with a passage through it and that rotates 90° from open to closed.

pneumatic actuator A valve actuator that uses pressurized gas as its energy source.

port *See* bore.

positive position The installation position of a multi-port valve such that the valve's flow control element is between the pressure inlet body openings and the body openings that are closed off, thereby assisting in sealing the valve.

pressure relief valve A valve that automatically opens at a set fluid pressure, allows fluid to discharge, and automatically recloses when the fluid pressure drops below the set pressure.

pressure seal joint A design of body–bonnet joint that uses the fluid pressure to compress a specially designed seal ring to form a tight seal and prevent external leakage.

quarter-turn valve A valve whose flow control element is rotated through 90° from open to closed; it can be a ball valve, a butterfly valve, or a plug valve.

quick-opening flow characteristic A flow characteristic in which there are large changes in the flow rate at the start of flow control element travel (valve begins to open), with progressively smaller changes with flow control element travel.

rack-and-pinion actuator A linear-to-rotary actuator that uses a gear pinion and rack to convert linear motion into rotary motion for use on quarter-turn valves.

reduced port A valve bore (port) found on gate valves and ball valves that is substantially smaller than full bore, approximately one pipe size on gate valves and 60% of full bore on ball valves.

regular port A valve bore (port) found on ball valves and plug valves that is smaller than full bore, approximately 75% to 90% of full bore on ball valves and 60% to 70% on plug valves.

regulator A type of control system or device that maintains the value of an output variable constant relative to a set input value.

relief valve A pressure relief valve designed for use with liquids.

resistance coefficient A measure of the resistance of a valve to fluid flow; it is determined by the geometry of the valve.

rising stem A stem that comes out of the valve (rises) as the valve is opened.

rotary actuator A valve actuator that directly produces rotary motion for use with quarter-turn valves.

safety relief valve A pressure relief valve that can be used with gases, liquids, and vapors.

Appendix 2

safety valve A pressure relief valve designed for use with gases and vapors.

Saunders valve A weir diaphragm valve.

scotch-yoke actuator A linear-to-rotary actuator that uses a scotch-yoke mechanism to convert linear motion to rotary motion for use on quarter-turn valves.

sealant A greaselike substance that is injected into lubricated plug valves to reduce operating force and to achieve a tight seal against internal leakage.

seat The portion of the valve body that the flow control element contacts to seal against internal leakage; it can be a separate part fastened in the body or can be integral with the body (a machined surface in the body).

seat ring A ring-shaped part that is fastened in a valve body to be a body seat; it is used in gate valves, globe valves, check valves, ball valves, and high-performance butterfly valves.

set pressure The pressure at which a pressure relief valve begins to open; it is established by adjusting the valve spring.

shaft The part of a ball valve, a butterfly valve, or a plug valve that turns the flow control element; on the plug valve it is usually integral with the plug.

shell The parts of a valve that hold in the fluid—the body, bonnet, cap, and so on. Shell parts are also called pressure-retaining parts.

short pattern valve A plug valve design that has a face-to-face or end-to-end dimension less than standard.

sleeved plug valve A plug valve design that has a nonmetallic lining on the body seating surfaces.

slide Common name for the gate of a knife gate valve.

slide valve *See* knife gate valve.

slurry A fluid consisting of small solid particles suspended in a liquid; for example, coal slurries and paper stock.

socket-weld end A valve end that is counter-bored to receive the connecting pipe, which is then fillet-welded to the valve.

soft seat A seat ring that is made of non-metallic material, usually plastic.

solder end A valve end that is counter-bored to receive copper tubing, which is then soldered to the valve.

solid-wedge gate A flow control element of a gate valve made of a single piece with no measures taken to introduce flexibility.

sour gas A natural gas containing a significant amount of hydrogen sulfide.

spherical seat A seat design of a globe valve in which the disc has a spherical seating surface that mates with a conical seat ring seating surface to produce line contact.

split-wedge gate A flow control element of a gate valve made of two separate pieces.

spring-return actuator A valve Actuator that compresses an internal spring when the actuator moves the flow control element away from its starting position (open or closed) and then uses the energy stored in the spring to move the flow control element back toward its starting position.

spur gear operator A gear operator that uses a spur gear set.

steam working pressure The maximum working pressure capability of a valve when used with steam, provided that the steam temperature does not exceed the maximum for the valve shell material; it is marked on the valve as S, SP, or SWP.

stem   The part of a gate valve, globe valve, or diaphragm valve that moves the flow control element.

stem bushing   *See* yoke nut.

stem nut   *See* yoke nut.

stop-check valve   A lift check valve design that also has a stem that can be used to hold the disc against its seat and have the valve function as a stop valve; also called a non-return valve.

stop valve   A valve used only for starting and stopping fluid flow.

stuffing box   The space in the bonnet of a valve that holds the packing used to seal against external leakage along the stem; also called a packing chamber.

swing check valve   A check valve design in which the flow control element rotates about an axis that is perpendicular to the fluid path and is outside the fluid path and the flow control element.

threaded end   A valve end that has female pipe threads cut into it to allow for screwing in male threaded connecting pipe.

throttling   The process of regulating the rate of fluid flow in a pipeline by moving the flow control element of a valve.

tilting-disc check valve   A check valve design in which the flow control element rotates about an axis that is perpendicular to the fluid path and is in the fluid path and passes through the flow control element.

translating stem valve   A gate, globe, or diaphragm valve. The valve's stem moves along its primary axis, and may also simultaneously rotate.

trim   The parts of a valve inside the shell that are "wetted" by the fluid, generally, the flow control element, its mover (stem or shaft), and the seal rings.

trunnion ball A flow control element of a ball valve that is held in position in the body by two integral, short-shaft extensions (trunnions) on the ball.

turbulent flow Fluid flow in which fluid "particles" move in irregular paths, and no two particles have similar paths; the motion of a particle at a given instance is not necessarily parallel to the overall direction of flow. Turbulent flow is not caused by changes in fluid direction or by obstacles in its path but is the result of fluid properties and velocity.

union joint A body–bonnet joint in which a female-threaded nut slides over the bonnet and screws onto a male-threaded body, clamping the bonnet in place.

valve A mechanical device used to control the flow of fluids in piping systems.

vapor A gas whose temperature and pressure are very near the liquid phase. Dry steam is considered a vapor because its state is normally near that of water. (Wet steam is a two-phase mixture of vapor and fluid particles.)

venturi port A valve bore (port) found on plug valves that is substantially smaller than a full port, approximately 40% to 50% of full bore.

wafer body A valve body that does not have "ends," but is held in place between two flanges by the flange studs.

wafer check valve A check valve design featuring two spring-loaded flow control elements and a wafer-style body.

water hammer A phenomenon that occurs when the velocity of a fluid in a pipeline is abruptly decreased (such as when a valve is rapidly closed). At the point of fluid velocity decrease there is a corresponding increase in fluid pressure that is reflected back upstream as a pressure wave, causing noise and vibration. If the initial pressure increase is great enough, the pipe may burst.

Water-Oil-Gas *See* cold working pressure. It is marked on the valve as WOG.

wedge A gate valve gate which is wedge shaped in cross-section.

wedge gate The gate of a valve that is wedge-shaped in cross-section for use with body seating surfaces that are inclined to the stem centerline; solid, flex, and split designs are available.

weir diaphragm valve A diaphragm valve featuring a dam, or weir, formed in the body across the fluid flow path. This design reduces diaphragm flexing, but increases flow resistance.

working pressure The pressure of the fluid passing through a valve measured at the valve inlet; fluid pressure varies as it passes through the valve and is lower at the valve outlet.

worm gear operator A gear operator that uses a worm gear set.

yoke An extension of a valve bonnet shaped like an inverted "Y"; the top of the yoke holds a yoke nut and the valve stem passes through it.

yoke bushing *See* yoke nut.

yoke nut An internally threaded nut held in a recess at the top of the yoke through which the valve stem passes. In gate valves and diaphragm valves the yoke nut is turned, and the stem is translated through it. In the globe valve the yoke nut is usually fixed and the stem is turned through it. It is also called a stem bushing, stem nut, or yoke bushing.

Y-pattern globe valve A globe valve design in which the stem is inclined at an angle of approximately 45° from the fluid path.

ISBN 0-8311-3077-6